Change and Motion: Calculus Made Clear 2nd Edition

Part I
Professor Michael Starbird

THE TEACHING COMPANY ®

PUBLISHED BY:

THE TEACHING COMPANY
4151 Lafayette Center Drive, Suite 100
Chantilly, Virginia 20151-1232
1-800-TEACH-12
Fax—703-378-3819
www.teach12.com

Copyright © The Teaching Company, 2006

ISBN 1-59803-232-1

Michael Starbird, Ph.D.

University Distinguished Teaching Professor of Mathematics,
The University of Texas at Austin

Michael Starbird is Professor of Mathematics and a University Distinguished Teaching Professor at The University of Texas at Austin. He received his B.A. degree from Pomona College in 1970 and his Ph.D. in mathematics from the University of Wisconsin, Madison, in 1974. That same year, he joined the faculty of the Department of Mathematics of The University of Texas at Austin, where he has stayed, except for leaves as a Visiting Member of the Institute for Advanced Study in Princeton, New Jersey; a Visiting Associate Professor at the University of California, San Diego; and a member of the technical staff at the Jet Propulsion Laboratory in Pasadena, California.

Professor Starbird served as Associate Dean in the College of Natural Sciences at The University of Texas at Austin from 1989 to 1997. He is a member of the Academy of Distinguished Teachers at UT. He has won many teaching awards, including the Mathematical Association of America's Deborah and Franklin Tepper Haimo Award for Distinguished College or University Teaching of Mathematics, which is awarded to three professors annually from among the 27,000 members of the MAA; a Minnie Stevens Piper Professorship, which is awarded each year to 10 professors from any subject at any college or university in the state of Texas; the inaugural award of the Dad's Association Centennial Teaching Fellowship; the Excellence Award from the Eyes of Texas, twice; the President's Associates Teaching Excellence Award; the Jean Holloway Award for Teaching Excellence, which is the oldest teaching award at UT and is presented to one professor each year; the Chad Oliver Plan II Teaching Award, which is student-selected and awarded each year to one professor in the Plan II liberal arts honors program; and the Friar Society Centennial Teaching Fellowship, which is awarded to one professor at UT annually and includes the largest monetary teaching prize given at UT. Also, in 1989, Professor Starbird was the Recreational Sports Super Racquets Champion.

The professor's mathematical research is in the field of topology. He recently served as a member-at-large of the Council of the American

Mathematical Society and on the national education committees of both the American Mathematical Society and the Mathematical Association of America.

Professor Starbird is interested in bringing authentic understanding of significant ideas in mathematics to people who are not necessarily mathematically oriented. He has developed and taught an acclaimed class that presents higher-level mathematics to liberal arts students. He wrote, with co-author Edward B. Burger, *The Heart of Mathematics: An Invitation to effective thinking*, which won a 2001 Robert W. Hamilton Book Award. Professors Burger and Starbird have also written a book that brings intriguing mathematical ideas to the public, entitled *Coincidences, Chaos, and All That Math Jazz: Making Light of Weighty Ideas*, published by W.W. Norton, 2005. Professor Starbird has produced three previous courses for The Teaching Company, the first edition of *Change and Motion: Calculus Made Clear*; *Meaning from Data: Statistics Made Clear*; and with collaborator Edward Burger, *The Joy of Thinking: The Beauty and Power of Classical Mathematical Ideas*. Professor Starbird loves to see real people find the intrigue and fascination that mathematics can bring.

Acknowledgments

I want to thank Alex Pekker for his excellent help with every aspect of this second edition of the calculus course. Alex collaborated with me substantially on the design of the whole course, on the examples and flow of the individual lectures, on the design of the graphics, and on the written materials. Thanks also to Professor Katherine Socha for her work on the first edition of this course and for her help during the post-production process of the second edition. Thanks to Alisha Reay, Pam Greer, Lucinda Robb, Noreen Nelson, and others from The Teaching Company not only for their excellent professional work during the production of this series of lectures but also for creating a supportive and enjoyable atmosphere in which to work. Thanks to my wife, Roberta Starbird, for her design and construction of several of the props. Finally, thanks to Roberta and my children, Talley and Bryn, for their special encouragement.

Table of Contents
Change and Motion: Calculus Made Clear
2nd Edition
Part I

Change and Motion: Calculus Made Clear
2nd Edition

Scope:

Twenty-five hundred years ago, the Greek philosopher Zeno watched an arrow speeding toward its target and framed one of the most productive paradoxes in the history of human thought. He posed the paradox of motion: Namely, at every moment, the arrow is in only one place, yet it moves. This paradox evokes questions about the infinite divisibility of position and time. Two millennia later, Zeno's paradox was resolved with the invention of calculus, one of the triumphs of the human intellect.

Calculus has been one of the most influential ideas in human history. Its impact on our daily lives is incalculable, even with calculus. Economics, population growth, traffic flow, money matters, electricity, baseball, cosmology, and many other topics are modeled and explained using the ideas and the language of calculus. Calculus is also a fascinating intellectual adventure that allows us to see our world differently.

The deep concepts of calculus can be understood without the technical background traditionally required in calculus courses. Indeed, frequently, the technicalities in calculus courses completely submerge the striking, salient insights that compose the true significance of the subject. The concepts and insights at the heart of calculus are absolutely meaningful, understandable, and accessible to all intelligent people—regardless of the level or age of their previous mathematical experience.

Calculus is the exploration of two ideas, both of which arise from a clear, commonsensical analysis of our everyday experience of motion: the *derivative* and the *integral*. After an introduction, the course begins with a discussion of a car driving down a road. As we discuss velocity and position, these two foundational concepts of calculus arise naturally, and their relationship to each other becomes clear and convincing. Calculus directly describes and deals with motion. But the ideas developed there also present us with a dynamic view of the world based on a clear analysis of change. That perspective lets us view even such static objects as circles in a dynamic way—growing by accretion of infinitely thin layers. The pervasive nature of change makes calculus extremely widely applicable.

The course proceeds by exploring the rich variations and applications of the two fundamental ideas of calculus. After the introduction in the setting of motion, we proceed to develop the concepts of calculus from several points of view. We see the ideas geometrically and graphically. We interpret calculus ideas in terms of familiar formulas for areas and volumes. We see how the ideas developed in the simple setting of a car moving in a straight line can be extended to apply to motion in space. Among the many variations of the concepts of calculus, we see how calculus describes the contours of mountains and other three-dimensional objects. Finally, we explore the use of calculus in describing the physical, biological, and even architectural worlds.

One of the bases for the power of calculus lies in the fact that many questions in many subjects are equivalent when viewed at the appropriate level of abstraction. That is, the mathematical structures that one creates to study and model motion are identical, mathematically, to the structures that model phenomena from biology to economics, from traffic flow to cosmology. By looking at the mathematics itself, we strip away the extraneous features of the questions and focus on the underlying relationships and structures that govern the behavior of the system in question. Calculus is the mathematical structure that lies at the core of a world of seemingly unrelated issues.

It is in the language of calculus that scientists describe what we know of physical reality and how we express that knowledge. The language of calculus contains its share of mathematical symbols and terminology. However, we will see that every calculus idea and symbol can be understood in English, not requiring "mathese." We will not eschew formulas altogether, but we will make clear that every equation is an English sentence that has a meaning in English, and we will deal with that meaning in English. Indeed, one of the principal goals of this series of lectures is to have viewers understand the concepts of calculus as meaningful ideas, not as the manipulation of meaningless symbols.

Our daily experience of life at the beginning of the third millennium contrasts markedly with life in the 17th century. Most of the differences emerged from technical advances that rely on calculus. We live differently now because we can manipulate and control nature better than we could 300 years ago. That practical, predictive understanding of the physical processes of nature is largely enabled by the power and perspective of calculus. Calculus not only provides specific tools that solve practical problems, but it also entails an intellectual perspective on how we analyze the world.

Calculus is all around us and is a landmark achievement of humans that can be enjoyed and appreciated by all.

Lecture One
Two Ideas, Vast Implications

Scope:

Calculus is the exploration of two ideas that arise from a clear, commonsensical analysis of everyday experience. But explorations of these ideas—the *derivative* and the *integral* —help us construct the very foundation of what we know of physical reality and how we express that knowledge. Many questions in many subjects are equivalent when viewed at the appropriate level of abstraction. That is, the mathematical structures that one creates to study and model motion are identical, mathematically, to the structures that model aspects of economics, population growth, traffic flow, fluid flow, electricity, baseball, planetary motion, and countless other topics. By looking at the mathematics itself, we strip away the extraneous features of the questions and focus on the underlying relationships and structures that govern and describe our world. Calculus has been one of the most effective conceptual tools in human history.

Outline

I. Calculus is all around us.

    A. When we're driving down a road and see where we are and how fast we are going…that's calculus.

    B. When we throw a baseball and see where it lands…that's calculus.

    C. When we see the planets and how they orbit around the Sun…that's calculus.

    D. When we lament the decline in the population of the spotted owl…that's calculus.

    E. When we analyze the stock market…that's calculus.

II. Calculus is an idea of enormous importance and historical impact.

    A. Calculus has been extremely effective in allowing people to bend nature to human purpose.

    B. In the 20th century, calculus has also become an essential tool for understanding social and biological sciences: It occurs every day in the description of economic trends, population growth, and medical treatments.

C. The physical world and how it works are described using calculus—its terms, its notation, its perspective.

D. To understand the history of the last 300 years, we must understand calculus. The technological developments in recent centuries are the story of this time, and many of those developments depend on calculus.

E. Why is calculus so effective? Because it resolves some basic issues associated with change and motion.

III. Twenty-five hundred years ago, the philosopher Zeno pointed out the paradoxical nature of motion. Zeno's paradoxes confront us with questions about motion. Calculus resolves these ancient conundrums.

    A. Two of Zeno's paradoxes of motion involve an arrow in flight.

        1. The first is the *arrow paradox*: If at every moment, the arrow is at a particular point, then at every moment, it is at rest at a point.

        2. The second is the *dichotomy paradox*: To reach its target, an arrow must first fly halfway, then half the remaining distance, then half the remaining distance, and so on, forever. Because it must move an infinite number of times, it will never reach the target.

    B. Looking at familiar occurrences afresh provokes insights and questions. Zeno's paradoxes have been extremely fruitful.

    C. Zeno's paradoxes bring up questions about infinity and instantaneous motion.

IV. In this course, we emphasize the ideas of calculus more than the mechanical side.

    A. But I must add that one of the reasons that calculus has been of such importance for these last 300 years is that it *can* be used in a mechanical way. It can be used by people who don't understand it. That's part of its power.

    B. Perhaps we think calculus is hard because the word *calculus* comes from the Greek word for *stones* (stones were used for reckoning in ancient times).

C. Calculus does have a fearsome reputation for being very hard, and part of the goal of this course is to help you see calculus in a different light.

D. In describing his college entrance examinations in his autobiography, Sir Winston Churchill says, "Further dim chambers lighted by sullen, sulphurous fires were reputed to contain a dragon called the 'Differential Calculus.' But this monster was beyond the bounds appointed by the Civil Service Commissioners who regulated this stage of Pilgrim's heavy journey." We will attempt to douse the dragon's fearsome fires.

E. Another reason calculus is considered so forbidding is the size of calculus textbooks. To students, a calculus book has 1,200 different pages. But to a professor, it has two ideas and lots of examples, applications, and variations.

V. Fortunately, the two fundamental ideas of calculus, called the *derivative* and the *integral*, come from everyday observations.

A. Calculus does not require complicated notation or vocabulary. It can be understood in English.

B. We will describe and define simply and understandably those two fundamental ideas in Lectures Two and Three. Both ideas will come about from analyzing a car moving down a straight road and just thinking very clearly about that scenario.

C. The viewer is not expected to have any sense whatsoever of the meanings of these ideas now. In fact, I hope these technical terms inspire, if anything, only a foreboding sense of impending terror. That sense will make the discovery that these ideas are commonsensical and even joyful, instead of terrifying, all the sweeter.

D. The derivative deals with how fast things are changing (instantaneous change).

E. The integral provides a dynamic view of the static world, showing fixed objects growing by accretion (the accumulation of small pieces).

F. We can even view apparently static things dynamically. For example, we can view the area of a square or the volume of a cube dynamically by thinking of it as growing rather than just being at its final size.

G. The derivative and the integral are connected by the *Fundamental Theorem of Calculus*, which we will discuss in Lecture Four.

H. Both of the fundamental ideas of calculus arise from a straightforward discussion of a car driving down a road, but both are applicable in many other settings.

VI. The history of calculus spans two and a half millennia.

   A. Pythagoras invented the Pythagorean Theorem in the 6th century B.C., and as we know, Zeno posed his paradoxes of motion in the 5th century B.C.

   B. In the 4th century B.C., Eudoxus developed the Method of Exhaustion, similar to the integral, to study volumes of objects.

   C. In about 300 B.C., Euclid invented the axiomatic method of geometry.

   D. In 225 B.C., Archimedes used calculus-like methods to find areas and volumes of geometric objects.

   E. For many centuries, other mathematicians developed ideas that were important prerequisites for the full development of calculus.

      1. In around 1600, Johannes Kepler and Galileo Galilei worked making mathematical formulas that described planetary motion.

      2. In 1629, Pierre de Fermat developed methods for finding *maxima* of values, a precursor to the idea of the derivative.

      3. In the 1630s, Bonaventura Cavalieri developed the "Method of Indivisibles," and later, René Descartes established the Cartesian Coordinate System, a connection between algebra and geometry.

   F. The two mathematicians whose names are associated with the invention or discovery of calculus are Isaac Newton and Gottfried Wilhelm von Leibniz. They independently developed calculus in the 1660s and 1670s.

G. From the time of the invention of calculus, other people contributed variations on the idea and developed applications of calculus in many areas of life.

    1. Johann and Jakob Bernoulli were two of eight Bernoullis who were involved in developing calculus.

    2. Leonhard Euler developed extensions of calculus, especially infinite series.

    3. Joseph Louis Lagrange worked on calculus variations, and Pierre Simon de Laplace worked on partial differential equations and applied calculus to probability theory.

    4. Jean Baptiste Joseph Fourier invented ways to approximate certain kinds of dependencies, and Augustin Louis Cauchy developed ideas about infinite series and tried to formalize the idea of *limit*.

    5. In the 1800s, Georg Friedrich Bernhard Riemann developed the modern definition of the *integral*, one of the two ideas of calculus.

    6. In the middle of the 1850s—about 185 years after Newton and Leibniz invented calculus—Karl Weierstrass formulated the rigorous definition of *limit* that we know today.

VII. Here is an overview of the lectures.

    A. In Lectures Two, Three, and Four, we will introduce the basic ideas of calculus in the context of a moving car and discuss the connection between those ideas.

    B. Then, we have a series of lectures describing the meaning of the derivative graphically, algebraically, and in many applications.

    C. Following that, we have a similar series of lectures showing the integral from graphical, algebraic, and application points of view.

    D. The last half of the course demonstrates the richness of these two ideas by showing examples of their extensions, variations, and applications.

    E. The purpose of these lectures is to explain clearly the concepts of calculus and to convince the viewer that calculus can be understood from simple scenarios.

1. Calculus is so effective because it deals with change and motion and allows us to view our world as dynamic rather than static.
2. Calculus provides a tool for measuring change, whether it is change in position, change in temperature, change in demand, or change in population.

F. Calculus is intrinsically intriguing and beautiful, as well as important.

G. Calculus is a crowning intellectual achievement of humanity that all intelligent people can appreciate, understand, and enjoy.

Readings:

Cajori, Florian. "History of Zeno's Arguments on Motion," *The American Mathematical Monthly*, Vol. 22, Nos. 1–9 (1915).

Churchill, Winston Spencer. *My Early Life: A Roving Commission.*

Any standard calculus textbook.

Questions to Consider:

1. Find things or ideas in your world that you usually view as complete and fixed and think about them dynamically. That is, view their current state as the result of a growing or changing process.

2. Explore the idea of function by describing dependent relationships between varying quantities that you see in everyday life. For example, in what way is the amount of money in a savings account dependent on interest rate and time? Or, for a less quantitative example, how is happiness a function of intellectual stimulation, exercise, rest, and other variables?

3. Think of a scenario from your daily life that interests you. Keep it in mind as you progress through the lectures—can calculus be applied to understand and analyze it?

Lecture One—Transcript
Two Ideas, Vast Implications

Welcome to calculus. It will be my great pleasure to guide you on a journey through a world of calculus during these 24 lectures. The tour begins at our doorsteps and takes us to the stars. Calculus is all around us every day of our lives. When we're driving down the road and we see where we are at every moment and we figure out how fast we're going, that's calculus. When we throw a baseball and see where it lands, that's calculus. When we see the planets and see how they orbit around the sun—calculus. When NASA sends a rocket ship to explore the solar system, and they make all these computations about the trajectories and how much fuel to burn in which direction, all of these involve calculus. When we turn on our TVs—calculus.

But, calculus is not restricted only to physical issues. When we lament the decline in the population of the spotted owl, that's calculus. When we analyze the stock market and we look at economic trends, that's calculus. Calculus comprises a collection of ideas that have had tremendous historical impact. And the reason is that calculus is enormously effective in allowing people to bend nature to human purpose. Much of the scientific description of our world is based on calculus; descriptions of motion, certainly; of electricity and magnetism; of sound waves or waterways—all of these involve calculus. But in addition to that, calculus is an essential tool for understanding social and biological sciences. It occurs every day when we describe economic trends, when we talk about population growth or decline, or medical treatments; all of the description are couched in terms of calculus. That is, its vocabulary, its notation, but most important, more important than any of those, are its ideas, its perspectives.

So, to understand the history of the last 300 years, we certainly must understand calculus. What is it that makes us experience life differently now from the way people understood it and experienced it 300 years ago? Well, most of the difference comes from technological developments, and most of those technological developments are the story of calculus; that is, the implications of calculus that allowed us to develop the technology that changed the way we experienced life. It would be extremely difficult to exaggerate the impact of calculus on our life experience.

So, that brings us to the question of why? Why is calculus so effective? And the answer is calculus resolves some basic issues associated with change and motion, just basic every day motion. If you think about every day motion, it turns out that that idea, just thinking about things moving, is a concept that is much trickier than we might at first realize. In fact, this was pointed out 2,500 years ago. The philosopher Zeno, in 450 B.C., proposed several paradoxes about motion, and let me first just say a word or two about what a paradox is. A paradox is an example where you look at some issue from two different points of view, and the two different points of view are both extremely persuasive. For each one you say, "That's the way to look at the thing." And then you discover that they are, in direct opposition to each other. So, when you have a paradox, it is pointing the way to a fundamental issue that needs to be resolved.

Well, Zeno's paradoxes were paradoxes about motion, and he forced us to confront some questions about basic motion that were not resolved for more than 2,000 years, and they were resolved by calculus. So, let me tell you about a couple of Zeno's paradoxes of motion. Two of them involve an arrow in flight. So, here was the idea of Zeno's paradox.

The first paradox was the following: you have an arrow shooting across the room, zooming across the room, and Zeno asked the following question, he said, "Okay. We see that the arrow is moving across the room, but let's ask the question, at one instant is it moving? At one instant it's in one place. So, in what sense is it, at that very instant, moving? This was a real puzzle. This was a real puzzle about motion. And, if you think about it, motion is one of these things that becomes more difficult to understand the more you think about it because of the nature of Zeno's paradox of the arrow. At every moment it is not moving; it's in one place. That was a big challenge for people historically, and it was not resolved until calculus.

Another example of Zeno's paradox was his paradox called the dichotomy paradox, and this was another paradox involving an arrow. This time somebody is standing over there and shoots an arrow right at me, and the arrow is coming at me. But, if I'm Zeno, I'm very calm. I'm calm because I think the following thoughts, I say, "Well, the arrow has to come halfway toward me, and then it has to come half of the remaining distance, and after that it has to come half of the remaining distance, and so on half and half and half." Therefore it has to accomplish infinitely many things; infinitely many traversals of distances. Consequently, it will never get to me and I can be perfectly sanguine in my future healthy by thinking about this paradox.

Well, of course, we know in both of these cases that reality is different. That is to say, we know the arrow is moving; we know the arrow will eventually strike its target. So, the question is, how can we resolve these issues that Zeno pointed out: instantaneous motion and the idea that there have to be infinite many traversals in order for the arrow to reach its target. Well, when we look at familiar issues, familiar occurrences, in new ways, we turn out to find all sorts of fruitful avenues for investigation. And in the case of Zeno's paradoxes, Zeno's paradoxes led to the calculus, among other things, which had tremendous impact, as we said, in many areas.

Zeno's paradoxes actually bring up issues about infinity, the case of infinitely many traversals before the arrow gets to the target and the question of infinity about motion. You see, the trouble with motion is that motion really refers to the basic idea that something in motion is at one place at one moment and in another place at another moment. Well, when you talk about instantaneous motion, you only have one moment to discuss. So, it's a question about dealing with zero divided by zero. The arrow has gone zero distance in zero amount of time, and that issue was a terrific conundrum that was not resolved until, in fact, not only—it was resolved by calculus, but it required further elaboration, even after calculus was invented, to really come to grips with an understanding of that. In fact, that took several hundred years after, or 200 years, after calculus was invented to really pin down that idea.

Well, in this course we're going to emphasize the ideas of calculus, the concepts of calculus, more than the mechanical side of calculus. But, I have to tell you that one of the main reasons that calculus has been of such importance during these last 300 years is that it can be used in a mechanical way. It can be used by people who don't really understand calculus. That's part of the power of calculus. And in fact, the word itself, the word *calculus*, refers to calculating; that is, calculating in this sort of mechanical way. And, by mechanical way, what I mean is this: We all learn how to multiply in elementary school, where we took two numbers and did something and we moved it down and we added things, and, well, most people, when they learn that mechanical way to do multiplication, they're learning a technique, but don't really understand why that technique leads to the answer that they're getting. So, it's in the same way that calculus has this calculating, mechanical quality to it, that a person can learn the mechanistic strategies by which calculus derives its implications without actually knowing what those, why they work, and what's involved in the conceptual background. And, I have to tell you, that my own students at The University of Texas, in my class, when they emerge from my class, they probably have that view of calculus, too, because when you take a calculus class, a lot of the time that you spend in the class is mastering these mechanical methods of doing the actual computations.

Well, in this set of lectures we're going to emphasize the ideas of calculus much more than the mechanical side. Actually, I think it's the mechanical side that gives calculus its reputation for being sort of a fearsome and maybe difficult and forbidding topic, but I have a theory about why calculus is often viewed as a very hard subject. Maybe it's because—and this is just a theory; I may be wrong—but I think it comes from the word. You see, calculus comes from the Greek word *calculi*, which is the Greek word meaning stones. See, stones were used for counting things in ancient Greek time. So, well, stones are hard, you see, and so that's why calculus—well, okay, maybe not. Okay, anyway. But, regardless of the reason, calculus certainly does have the reputation for being extremely hard and sort of fierce, and part of the goal of this course is to, of course, make it seen in a totally different light. Calculus can be our friend.

But, you know, I teach mathematics and I teach calculus and so I often hear about the fright and the pain associated with mathematics in general, or calculus in particular, like when I'm in an airplane and I sit next to somebody and I divulge that I teach mathematics or I teach calculus. Well, they first tend to move away; that's the first thing that happens. They've sort of been afraid of it. But, one of—a particularly delightful example of this representation of the fear and loathing of mathematics and calculus in general came about in Sir Winston Churchill's autobiography, where he was writing about his early life in his autobiography called *My Early Life, A Roving Commission*. And in Chapter Three of this, he talked about taking examinations which he had to pass in order to enter the military academy. He had trouble passing these examinations—which, of course, is always astounded me that Winston Churchill, of all people, would have trouble passing examinations. But, that's just the way it was. So, in this Chapter Three he talks about mathematics and how he had to learn mathematics to pass this examination. So, he says the following: "All my life, from time to time, I've had to get up disagreeable subjects at short notice; but I consider my triumph, moral and technical, was in learning mathematics in six months." So that was, first of all, was pretty good. He goes on to describe his recollections of these days. "When I look back upon those care-laded months, their prominent features rise from the abyss of memory. We were arrived in an *Alice in Wonderland* world at the portals of which stood a quadratic equation." Here's where he gets to calculus. He said "Further dim chambers, lighted by sullen, sulphurous fires, were reputed to contain a dragon called the 'Differential Calculus.' But this monster was beyond the bounds appointed by the Civil Service Commissioners who regulated this stage of pilgrim's heavy journey." Isn't this great? So, calculus has this reputation of being a real monster; a dragon with a fiery breath. But, anyway, we will attempt to quench these fearsome fires. By the way, I must say, reading this little excerpt from Churchill—Churchill, of course, is amazing, having written 52 volumes of history; he won the Nobel Prize in literature; so, it is delightful to read about Churchill, and particularly this book about his early life.

But I think there are other reasons why calculus is viewed as a difficult subject and sort of forbidding, and I have to admit that I think mathematicians will not win the prize for using psychology that calms and sooths the soul. I think we're not going to win that prize because I think another reason that calculus is so forbidding is the simple size of calculus textbooks. Now, I brought a sample here—this is a calculus textbook—and, first of all, it does have a lot of benefits, not only mental, but also physical because people—you learn to heft heavy weights and so you get strong in that way, too. But, this is a typical calculus textbook; it has more than 1,300 pages. And, of course, to a student—and they're very big pages and sort of forbidding—that, to a student, these 1,300 pages are all different, but to a professor it has two ideas and 1,298 pages of examples, applications, and variations. The range and the richness of the implications of those two ideas show the power of those two insights. The exploitation of those two fundamental ideas leads to this whole world of calculus. Fortunately for us, and for this course, those two fundamental concepts of calculus—which, by the way, are called the derivative and the integral—both of those ideas come from every day observations. Calculus does not require a complicated vocabulary or notation to understand it really authentically. It can be understood in English; and, for the most part, we will be talking, in this course, in English about the ideas of calculus.

Okay, so calculus is the study of these two basic ideas: the *derivative* and the *integral*. Now, we're going to describe them and we're going to clearly say them in an understandable way in Lectures Two and Three respectively; derivative Lecture Two and integral Lecture Three. These ideas both will come about by analyzing a very simple situation of a car driving down a straight road; and then thinking very clearly about the implication of that idea and how it is that we need to think about such questions as Zeno's paradox of instantaneous motion. What are we going to mean by that? Now, let me make absolutely clear, that I'm not expecting any viewer of this course to have any sense whatsoever about the meaning of these ideas. In fact, in a way, I sort of hope the technical terms "derivative" and "integral" inspire in you, maybe, something that's foreboding, maybe a sense of impending terror, because that way when you find out that, in fact, these ideas are accessible and enjoyable, it will be all the sweeter. So, you see, this is good.

Well, the derivative deals with instantaneous change—how fast things move and change; and the integral gives us a dynamic view of the world, even of the static world. It talks about the accumulation of little things that accumulate into a big thing, and that's what I love about calculus. That it can take even the—it gives us a dynamic view of the world. That even things such as the area of a square or the volume of a cube, all of these things that seem so static can be viewed as growing into the size of the cube or growing into the size of a ball, and looking at that perspective allows us to see the world in a different way.

Well, the derivative and the integral turn out to be connected to each other, and they're connected by the fundamental theorem of calculus; and, we'll talk about that in Lecture Four. But, as I said before, both of these ideas arise from the discussion of a car moving down a road, but the same ideas that we get from analyzing this car moving down the road are applicable in these many, many other settings. That's the strength and the power of mathematics in general, and of these two ideas, derivative and integral, in particular.

Now, we're not going to, in this course, we not going to take the shortcut of over-simplifying. Einstein had a quote on that; it went something like, "I want things presented to me as simply as possible, but not simpler." And in this course, we're going to present things that really give you the authentic understanding of the ideas that make up calculus.

But the two ideas that come from calculus that really make up the foundations of calculus, did not appear by magic. The history of calculus spans two and a half millennia. From the time of Zeno, it was more than 2,000 years that elapsed before calculus became fully developed; and during that time there was a steady progress of mathematical ideas that needed to be created before we can understood calculus the way we know it now. So, I thought that we might take a few minutes here and go through and look at some of the people who participated in this long development and to hear a little about them.

So we should actually begin a little bit before Zeno, in the 6th century, Pythagoras started his famous school of mathematics and, of course, he proved the Pythagorean Theorem. Then Zeno came along in the middle of the 5th century B.C. In the middle of the 4th century B.C., there was a Greek mathematician by the name of Eudoxus, who developed a method of exhaustion that was very similar to the integral, one of the basic ideas of calculus. In fact, we'll see an example of Eudoxus's method in Lecture Number Nine.

In about the year 300 B.C., Euclid invented the axiomatic method of geometry and wrote maybe the most famous mathematics book ever written, *The Elements*. In 225 B.C., Archimedes used a method of exhaustion to develop formulas for areas and volumes of geometrical figures, and we'll see a great example of that that, in fact, involves his lever, in Lecture Number Nine.

Well, let's skip ahead a couple thousand years and get to the year 1600. There were some contributions in between, but it really got going again in the year 1600 when we have Johannes Kepler and Galileo Galilei working on motion, that planetary motion and describing the motion of objects just in the world, and making mathematical formulas that would describe that motion.

In 1629, Pierre de Fermat developed methods for finding maxima of values; and, in fact, that idea was very close to the idea of the derivative and we'll see it applied in our lecture about optimization. So, he was very close to being an inventor of calculus, but not quite.

In the 1630s, Bonaventura Cavalieri developed a method that he called the "method of indivisibles," and we'll see an example of that in a later lecture. René Descartes was famous for his Cartesian coordinate system, made a connection between algebra and geometry.

In 1669, Isaac Barrow, who was Newton's teacher, gave up his chair in mathematics for his famous student, Newton, and I thought this was such a nice thing to do until I learned that actually he had gotten a better job and went into a different line of work.

But finally, after all this, we come to the two people whose names are most closely associated with the development of calculus, and these are Isaac Newton and Gottfried Wilhelm von Leibniz. These are two people who are most associated with calculus, and we might say they invented calculus or discovered calculus; but, actually, all of the developments of calculus were definitely incremental. And the fact that both of them independently invented calculus at nearly the same time shows that the ideas were in the air. In fact, Isaac Barrow, Newton's teacher, remember, he actually expressed the fundamental theorem of calculus in one some of his lectures, but didn't appreciate the significance of calculus. He wrote it down, but didn't appreciate the significance. It took Newton and Leibniz to really systematize calculus, which was what we come to think of as the core of it.

Well, from the time that the calculus was actually invented in the late 17th century, it still took many years for calculus to be really understood; and many people developed variations to the idea and applied it to many applications in life, from physics, economics, probability—all sorts of things, biology. Johannes and Jakob Bernoulli were two of eight Bernoullis that were involved in developing calculus and applying it in Europe. Leonhard Euler developed a tremendous amount of mathematics, including many applications and extensions of calculus, especially infinite series. Joseph Louis Lagrange worked on the calculus variations. Pierre Simon de Laplace worked on partial differential equations and applied calculus to probability theory. John Baptiste Joseph Fourier invented ways to approximate certain kinds of dependencies, certain kinds of descriptions of things, in terms of what are called circular functions, like sines and cosines. And, by the way, this idea of taking a path, such as the path of a planet, and describing it by a collection of circles, really came from the old idea of using epicycles to try to describe the elliptical orbits of planets. So, he by adding a lot of these circular motions, you can approximate other kinds of curves. Augustin Louis Cauchy developed ideas about infinite series and tried to formalize the idea of limit. We'll find that the idea of a limit comes up in both the derivative and the integral, and it is one of the challenges to calculus that neither Newton nor Leibniz really were able to master. It took another 180 years before people really could pin down that idea of limit, which is when things, numbers, are converging toward one number. But we'll talk about that much more in future lectures.

In the 1800s, Georg Friedrich Bernhard Riemann developed the modern definition of the integral, which is one of those two ideas of calculus; and, in the middle of the 1850s, Karl Weierstrass formulated, finally, the rigorous definition of limit that occurs in this calculus book and we use today. Notice, again, that this 1850 is 185 years since Newton first developed calculus in 1665.

So, it took a long time to understand calculus, even after it was invented. So, if you don't understand calculus, you can take some solace in the fact that it took humanity 2,000 years to develop it and, even after it was developed, it took several hundred years to understand. It takes everybody a long time to understand it. And, in fact, essentially nobody understands calculus the first time they take it. I didn't understand calculus the first time I took it and among my friends who teach mathematics, they didn't understand calculus the first time they took it. Almost everybody—in fact, for most of us who teach calculus, the time we understood calculus was the time we taught it, and I recommend this as a method for learning anything.

Well, this then, gives a sense of the history of calculus. I'd like to take the remaining part of this lecture to just tell you the structure of the upcoming lectures, the structure of the course.

So, we'll begin in Lectures Two, Three, and Four comprise a collection of lectures that really present the fundamental ideas of calculus. In Lecture Two we introduce the idea of the derivative and say what the definition is and what it means. Then, in Lecture Three, we do the same thing to the integral and say what the definition of the integral is, how it comes about. And, then, in Lecture Four, we introduce the fundamental theorem of calculus, which connects the two. And, if I have done my job correctly, you will find that Lecture Four is sort of a letdown in that it is obvious that the derivative and the integral are two sides of the same coin; that they are related to each other. But, that was an insight that took some time, historically, to clarify.

Well, after we get through Lecture Four, then we proceed to investigate each of these fundamental ideas in more detail. The next lectures are a sequence of lectures about the derivative, and in the collection of lectures about the derivative, we're going to look at the derivative and see that, in fact, although we had presented the idea of the derivative in terms of motion, of a car moving on a straight line, we'll see that that same abstract procedure that's the derivative also tells us other things. It tells about—it can be interpreted in terms of graphs of functions, which we'll talk about and introduce in later lectures. But it says the same procedure that tells us the instantaneous velocity of a car also tells us the slope of a tangent line of a curve, and that kind of connection is an interesting one that gives the power to the ideas of the derivative.

Similarly, we're going to take the derivative and look at it from the point of view of its algebraic manifestations. One of the properties that makes the derivative so potent, and the integral, is that you can do it algebraically, and that's this mechanical side that students view as—the most common part that they deal with most is learning how to manipulate the algebra. So, we'll see the derivative then physically, that is, with the car moving; and then graphically; and then algebraically; and, then, we'll see it applied to different application areas such as volume, formulas for the volumes of objects—all of these things have manifestations about the derivative.

After that, we turn to the integral and we have a similar sequence of lectures that present the integral in these terms. That is to say, graphically, algebraically, and in terms of their applications to many different areas. So by taking these fundamental ideas and viewing them in different ways, that will show the richness of these themes. Then, the last half of this course will demonstrate the richness of these two ideas by showing lots of examples of their extensions, their variations, and their applications.

Well the purpose of these lectures is to explain clearly the concepts of calculus and to convince you that calculus can be understood from simple scenarios. Calculus is so effective because it deals with change and motion, and it allows us to view our world as a dynamic rather than just a static place. Calculus provides a tool for measuring change, whether it's a change in position, change in temperature, change in demand, or change in population. But, in addition to that, I like to think that calculus is intrinsically intriguing and beautiful, as well as just being merely important. So, calculus is a crowning intellectual achievement of humanity that all intelligent people can appreciate, enjoy, and understand. I look forward to exploring calculus with you during the next 23 lectures. Bye for now.

Lecture Two
Stop Sign Crime—
The First Idea of Calculus—The Derivative

Scope:

Change is a fundamental feature of our world: temperature, pressure, the stock market, the population—all change. But the most basic example of change is motion—a change in position with respect to time. We will start with a simple example of motion as our vehicle for developing an effective way to analyze change. Specifically, suppose we run a stop sign, but in preparation for potential citations, we have a camera take a picture of our car neatly lined up with the stop sign at the exact instant that we were there. We show this photograph to the officer and ask to have the ticket dismissed, presenting the photograph as evidence. The officer responds by analyzing our motion in a persuasive way that illustrates the first of the two fundamental ideas of calculus—the *derivative*. We get the ticket but can take some solace in resolving one of Zeno's paradoxes.

Outline

I. Calculus has two fundamental ideas (called the *derivative* and the *integral*)—one centered on a method for analyzing change; the other, on a method of combining pieces to get the whole.

  A. Both of the fundamental concepts of calculus arise from analyzing simple situations, such as a car moving down a straight road.

  B. This lecture presents an everyday scenario that leads to one of the two ideas of calculus—the derivative.

II. The following stop-sign scenario is a modern-day enactment of one of Zeno's paradoxes of motion.

  A. Let us suppose we have a car driving on a road, and there is a mileage marker at every point along the road. Such a simple scenario can be represented in a graph.

    1. The horizontal axis is the time axis.

    2. The vertical axis tells us the position of the car at each moment of time.

3. For the sake of arithmetic simplicity, we will talk about measuring the velocity (speed) of the car in miles per minute. Therefore, the vertical axis of our graph is in miles and the horizontal axis is in minutes.

B. Suppose Zeno is driving this car, and he goes through a stop sign without slowing down.

C. Soon thereafter, he is pulled over by Officers Newton and Leibniz. (The corny names will make memorable how the roles in this drama relate to Zeno's paradox and the invention of calculus.)

D. The driver, Zeno, protests by showing a still picture of his car exactly at the stop sign at the exact moment, 1 minute after the hour, when he is supposed to have been running the stop sign. On this street, time is measured by stating the minutes only.

E. Zeno claims that there could be no violation because the car was in one place at that moment.

III. Officers Newton and Leibniz produce additional evidence.

A. The officers produce a still photo of the car at 2 minutes after the hour clearly showing the car 1 mile beyond the stop sign.

B. Zeno argues, "So what? At 1 minute, I was stopped at the stop sign."

C. Newton: "But you must admit your average velocity between 1 and 2 minutes was 1 mile per minute."
  1. To compute the average velocity during any interval of time, you need to know the position of the car at the beginning, the position at the end, and the amount of time that passed.
  2. The average velocity is change in position divided by change in time, that is, how far you went divided by how long it took.
  3. The average velocity does not rely on what happened between those two moments—just on where the car is at the beginning and at the end.

D. Again, Zeno says, "So what?"

E. Officers Newton and Leibniz produce an infinite amount of additional evidence, all incriminating. They note that Zeno was at the 1.1 mile marker at 1.1 minute, at the 1.01 mile marker at 1.01 minute, and so on, all of it proving that Zeno's velocity was 1 mile per minute.

IV. The idea of instantaneous velocity is the result of an infinite amount of data.

    A. All the evidence is that, on every even incredibly tiny interval of time, the average velocity was 1 mile per minute.

    B. The cumulative effect of all the evidence—an infinite number of intervals—leads to the idea of instantaneous velocity.

V. Knowing the position of a car at every moment allows us to compute the velocity at every moment. This can be illustrated with a car whose velocity is increasing.

    A. Let's now consider the example where at every time, measured in minutes and denoted by the letter t, we are at mileage marker t^2 miles. For example, at time 1, the car is at position 1, but at time 2, it is at position 4 (2^2), and at time 0.5, it is at position 0.25 (0.5^2).

    B. If we know where the car is at every time during an hour, we can tell how fast it was going at any selected moment by doing the infinite process of finding instantaneous velocity.

    C. Let's apply that infinite process to this moving car at several different times t.

        1. First, if time $t = 2$ minutes, the position of the car is $2^2 = 4$. Then, if time $t = 1$ minute, the position of the car is $1^2 = 1$. Subtracting 1 from 4 and dividing by 1 minute, we have 3 miles per minute. In other words, by looking at where the car was 1 minute after the 1-minute mark, we find that the average velocity was 3 miles per minute.

        2. However, when we look at shorter intervals of time, we get a different story, as shown in the chart below. We will find the positions of the car at various nearby times, such as 1.1, 1.01, 1.001, and 0.99 minutes, and compute our average velocity between time 1 and those times.

Position: $p(t) = t^2$		
Initial Time	**Final Time**	**Average Velocity (mi/min)**
1	2	3
1	1.1	2.1
1	1.01	2.01
1	1.001	2.001
0.999	1	1.999
Instantaneous velocity at $t = 1$ is 2 mi/min		

3. The whole collection of average velocities leads us to conclude that the instantaneous velocity is 2 miles per minute.

D. This process of taking smaller and smaller intervals of time to arrive at the instantaneous velocity is called a *limit*; the instantaneous velocity is the limit of the average velocities as the intervals of time get smaller and smaller.

E. The infinite process used to find the velocity at each time is the *derivative*. The derivative gives the instantaneous velocity at any given moment using this infinite process, if we know the position of the car at each moment.

F. If we look at the same question for other times, 0.7 minutes, for example, we have the following results:

Position: $p(t) = t^2$		
Initial Time	**Final Time**	**Average Velocity (mi/min)**
0.7	1.7	2.4
0.7	0.8	1.5
0.7	0.71	1.41
0.7	0.701	1.401
0.699	0.7	1.399
Instantaneous velocity at $t = 0.7$ is 1.4 mi/min		

G. If we look at the same question for other times, such as 1.4, 2, or 3 minutes, we have similar results for the instantaneous velocity.

H. We have been computing instantaneous velocities at various times for a car that is moving in such a manner that at every time t minutes, the car is at mileage marker t^2 miles. If we look at a chart of all our examples of instantaneous velocity, we see a pattern that indicates that at every time t, the velocity is $2t$ miles per minute.

Position: $p(t) = t^2$	
Time (min)	**Instantaneous Velocity (mi/min)**
0.7	1.4
1	2
1.4	2.8
2	4
3	6
Instantaneous velocity: $v(t) = 2t$	

VI. The first idea of calculus, the derivative, quantifies the idea of instantaneous velocity.

    A. We have taken a simple, everyday scenario (a moving car) and developed a simple, though infinite, process that made sense of the intuitive idea of motion or velocity at an instant.

    B. We now have an equation that tells us the instantaneous velocity at each moment of time.

        1. The derivative of a function $p(t)$ at time t is obtained by computing $\dfrac{p(t+\Delta t) - p(t)}{\Delta t}$, where Δt is a small increment of time, and then seeing what number those values approach as Δt becomes increasingly small.

        2. The single number to which those values approach is called the limit: $\dfrac{p(t+\Delta t) - p(t)}{\Delta t}$ as Δt approaches 0.

Readings:

Any standard calculus textbook, section introducing the derivative.

Questions to Consider:

1. Suppose a rocket is traveling along a road, and at each time t minutes its position is t^3 miles from where it started. Use the derivative process (and a calculator) to guess the instantaneous

velocity of the rocket at various times, such as 1 minute, 2 minutes, and 3 minutes after starting.

2. Would it make sense to view position and time as discrete quantities having a little width to them? Then, a moving object could pause for an instant of time at one "point" before moving on. Zeno's fourth paradox of motion treats this situation when he considers three objects on parallel tracks, one moving right, one moving left, and one fixed. The relative velocities of these moving objects present a challenge. What is it?

Lecture Two—Transcript

Stop Sign Crime—
The First Idea of Calculus—The Derivative

Welcome back to *Change and Motion: Calculus Made Clear*. In this lecture, the word for the day is *derivative*. Today the whole purpose of this lecture is to introduce the idea of derivative and to explain what it means. So, it's a very clear goal for the lecture. Remember, calculus is based on two ideas; it's the exploration of two ideas. One of them is the derivative, which we'll meet today, and the other one is the integral, which we'll meet in lecture number three. And both of them come from the same scenario of a car driving down a straight road, and so, our challenge today is to analyze this very simple situation of a car moving down a straight road; and, from it, deriving the definition and the concept of the derivative.

So, let's just get started right away with the following scenario: Suppose that we have a car that's driving along a road. And so, here, we have a car, you see, and it's driving a long a straight road like this; at every moment of time the car is in some place along the road, and this road is amazingly well marked. You know how there are mileage markers on some roads? Well, this is a road that has a mileage marker at every single point on the road, so you know exactly where you are at every single moment. So, at every moment of time, the car is at some position on the road, you see? So, this is a basic idea; and, one thing that I wanted to say before we develop the scenario of the derivative is the idea of how can we represent a simple scenario such as a car moving on a road in a way that is useful to describe. Well, one way is to look at a graph that captures the idea of a car moving in a road. The way that a graph works is very simple. We have a horizontal axis, which is the time axis, and then we have a vertical axis that tells us the position of the car at each moment of time. So, if we have a car that's moving, for example, just like this on our road, and it's sort of a steady kind of a way, then that would correspond to the car's position changing, and it could be recorded that at each moment of time it is in a certain position. And, so, every dot on this graph represents the position of a car moving on a road.

Now, I do want to say one thing about this: That, in all of the lectures today and tomorrow and the next day, we're going to be talking about this car moving on a road, and for purposes of arithmetic simplicity, we will be talking about measuring the speed of the car in miles per minute rather than miles per hour, and the reason is that 1 mile per minute is a smaller number than 60 miles per hour. So, it's just computationally easier. So, in our graph here, the vertical axis will be in miles and the horizontal axis will be in minutes.

Okay, now, here is the scenario that we're going to discuss first in our situation of a car moving down the road. Here's the car; and here is, as you see, a stop sign. Now, here's the way the car goes; it just goes like this. I guess the stop sign should really be like this. Now, you notice something about that car moving down the road; the car moved down the road—notice, it didn't stop. It just went straight through the stop sign. Now this, by the way, is going to illustrate Zeno's paradox, his arrow paradox, but in the modern terms of a car moving down the road.

Now so here's what happened in our scenario: This car moved down the road, it just saw the stop sign, and then just drove right on through. That was what the car did. A little later, the car, as it was driving down the road, it heard a siren in the background and the policemen pulled up and stopped the car and said, "You just went through that stop sign. What do you think you're doing? You didn't even slow down? You just went right through the stop sign." And the driver of the car, though, it turns out that the driver of the car—by the way, the driver of the car's name is Zeno; that's the driver of the car—was prepared for this possibility for being pulled over, and said, "No, no, officer, officers, that's not true. You say that I went through the stop sign; but, in fact, I have a photograph that proves that I was just at the stop sign at the moment that you say I was moving through it;" and so Zeno had a picture of the car right at the stop sign. See? A still photograph of the car right at the stop sign which he had prepared in advance to be taken, and he took that and it was sent electronically to him in this car, and he said, "Here is this photograph of the car at the line of the stop sign. What do you mean by claiming that I was moving at that time? I wasn't moving. Look at this still picture. Any jury in its right mind could not possibly convict me of going through that stop sign when I have this evidence that I was in one place at one time."

Well, at that point, Zeno looks at the little nametags of the officers, and one of the officer's names is Newton and the other officer's named Leibniz. Now, Zeno, if he had lived a few thousand years later, would be a little bit nervous at this point because it turned out that Newton and Leibniz, these officers, say, "Ah, but we have some additional evidence about this case; we have additional evidence that's going to illustrate why you were moving at the time you claim you were stopped at the stop sign; why you were moving at the stop sign." So, they say, "Look. You have this photograph of you right at the stop sign at a given time, but we have a photograph of you one minute later one mile down the road. See? And, so, we can compute how fast you were going during that minute, and here's the way we do it." You see, it's very easy to compute how fast a car is going between two moments of time because you simply look at the position that the car was in at the last time and the position the car was in at the first time, subtract the two to get the distance traveled— so, if this is the position of the car at the first time, and this was at the last time, then that difference in distance is the distance traveled; and, then, dividing by the amount of time it took, in this case one minute, would give us the average velocity in that the car was traveling during those two instances of time. And that's the nature of motion. The nature of motion—the most fundamental aspect of motion is that the car is at one place at one instant and it's at another place at another instant.

Now, let's be specific about our times so we can make this computation exact. Our time is being measured in minutes, and, if you wish, you may think of these as minutes after an hour. The car was at the stop sign at precisely 1 minute, your watch was reading exactly 1 minute; that's where the car was. One minute later, the car was 1 mile down the road. So, we have the photograph of the car at 1 minute past the hour, and here's the photograph of the car at 2 minutes past the hour. The position of the car means where the car was against the mileage marker running along the side of the road. Now, these markers, by the way, don't correspond.

This is a ruler; don't look at the ruler for the mileage markers. We're going to say— I'll put the ruler down—that at the time that the

$$\text{Average speed} = \frac{p(2) - p(1)}{2 - 1} = \frac{2 - 1}{1} = 1 \text{mi/min}$$

car was at the stop sign, the stop sign was at the 1-mile position on this straight road. And, then, 1 minute later, it was at the 2-mile position on the straight road. In other words, you look out your window and that's what you see, the 2-mile marker. So, here we can compute the average speed, or the average velocity, during that time was the position of time 2 minus the position of time 1, divided by the elapsed time, which was—the time 2 minutes was what time the person ended up 1 mile down the road; and then minus 1 minute, that was the time that the person was at the stop sign; and this division, $\frac{2-1}{1}$, is 1 mile per minute. And that's the average time, the average velocity, of the car during that minute.

Well, of course, Zeno says, "So what? So what? What do you mean to say that I was going through the stop sign because all of the evidence that you have is where I was at the time of the

$$\text{Average speed} = \frac{p(1.1) - p(1)}{1.1 - 1} = \frac{1.1 - 1}{1} = 1 \text{mi/min}$$

stop sign and where I was a minute later. But how do you know that I wasn't stopped at the stop sign and then I just speeded up to get a mile down the road in the minute? It doesn't tell me anything about what I was doing at the instant that you claim I went through the stop sign."

Well, the two officers, Leibniz and Newton, say, "We have additional evidence. We had other photographs taken, and the other photographs we had taken were photographs that took place at exactly 1.1 minute—the time was 1.1 minute—we took a photograph, and where was the car? It was at the position 1.1 mile marker." Well, and so what does that say? That says that the position at time 1.1 was 1.1; that's where the car was at time 1.1. At time 1, the car was at position 1. The difference between those two is .1. So, the car traveled .1 mile in the elapsed time. The elapsed time was 1.1, that was the time that we measured at the second position minus the initial time, 1; elapsed time of .1; dividing the two gives us a value of 1 mile per minute.

Zeno says, "Look. That still doesn't prove it. That just says where I was at the beginning and the end of this interval; it doesn't tell me how fast I was going at the instant you claimed I was moving through the stop sign."

So, Newton and Leibniz say, "Wait a minute; we have more evidence. Look at time 1.01; where were you? You were at the position 1.01; the elapsed time was .01 minutes; dividing it out, we see that it was the same story—you were going an average of 1 mile per minute when measured by these two moments of time that were very, very close to each other." And now then they say, "Look. We have more evidence. We have two pictures—one at time 1.001; where were you? You were at position 1.001. So, you traveled a total of .001 miles in .001 minutes; which, once again, is telling the story of 1 mile per minute." And in fact, Newton and Leibniz have infinite amount of information to give to the jury. They have the information of every single instant of time of where the car was before and after the time message number 1, and each time it told the same story. The story was that the car was going at 1 mile per minute. Even taking a photograph of where the car was immediately before the 1-minute time, at the .999 minute on your watch, the car was at position .999 on the road. Since all of that evidence, an infinite amount of evidence, all is telling the same story, we conclude that the instantaneous velocity at the 1-minute mark was exactly 1 mile per minute.

Okay, by the way, you may say this was a long and tedious story, and I have to tell you, this is a bit of an expository challenge because the whole concept of the derivative is, by its very nature, repetitive and similar. We're doing the same thing, subtracting the position at one moment minus the position at another moment and dividing by the elapsed time, and we're doing it over and over and over again, infinitely many times, and we're seeing that it's telling the same story; and that's what we're going to define as the instantaneous velocity of the car at time 1 because all of these measurements are telling us that the car was going at average speeds, average velocities, of 1 mile per hour, and therefore that is going to be declared to be the instantan

$$\text{Average speed} = \frac{p(1.001) - p(1)}{1.001 - 1} = \frac{1.001 - 1}{.001} = 1 \text{ mi/min}$$

eous velocity at that time.

Notice, in order to talk about the instantaneous velocity, we needed information about the position of the car at all times nearby, and that was the thing that allowed us to make a decision about the instantaneous velocity at that moment. So, it was not just the information about where the car was at that moment, as Zeno would like it to have been for his defense; that photograph of him correctly at the stop sign at that time, that's not sufficient. In fact, you needed to know where the car was at all times nearby.

Now, you may say, well, okay, this is the first thing you'd think of—maybe not; I don't know if you think that way or not. But, I wanted to read to you some statement from Aristotle, who thought about Zeno's paradoxes and tried to shed light on them, and here is what Aristotle had to say about motion because instantaneous motion was a real conceptual challenge to people for many millennia. So, here's what he said. Aristotle was trying to give a definition of motion; he said the following: "Motion is the fulfillment of what exists potentially insofar as it exists potentially." And he goes on to say, here's a further remark on motion, "We can define motion as the fulfillment of the moveable qua moveable."

Now, as you know, I am a teacher, and I sometimes get papers from students that say things like that, and I always wonder did they feel that they were actually saying something meaningful? Maybe Aristotle felt he was shedding light on this issue, but it certainly is difficult to get anything from it. But, the point is, that we have had to do an infinite process to come up with a single number.

Now, there was a defect to the example that I just gave, and the defect was that in all instances, for every single pair of instances of time, we got the same answer. So, all the evidence was leading to the conclusion that the car was going at 1 mile per minute; every single time it said 1 mile per minute. Let's do another example.

Here's an example where the car is moving according to another scenario. So, here we have our car; it's moving on our, once again, on our well-traveled road, but this time the car, instead of going at a steady velocity, the car is speeding up as it goes. It's speeding up according to a very well-defined notion, and that is the following: That at every moment of time, you look at your watch and you say that the car is going to be at the position that's the square of the time that's on your watch. So at time 1 it is at position 1, as before; but at time 2, it is at 2×2 at position 4; and at time .5, it's at position .25. Follow me? So, that it's speeding up as it goes; speeding up as it goes. So that is the position of the car. [This is represented by the equation $p(t) = p^2$.]

Let's see if we can undertake exactly the same question about what the instantaneous velocity of the car is at time 1. Let's undertake the same thing. Now I know it's laborious, but it's important because this is one of the two basic ideas of calculus. So, here we go.

Suppose that you looked at the

$$\text{Average speed} = \frac{p(1.1) - p(1)}{1.1 - 1} = \frac{1.21 - 1}{.1} = 2.1$$

position of the car at time 2. Well, the position is given by taking the time and squaring it, multiplying it by itself, so that's 4. So, the position at time 1 is given by squaring that, so it's at position 1, so the difference is 4 minus 1, that is, the car traveled 3 miles during that 1 minute of elapsed time, between time 2 and time 1; dividing 3 by 1, we get 3 miles per minute. So, by looking at where the car was 1 minute after the 1-minute mark, and we're undertaking to understand what the instantaneous speed was at time 1 minute—well, by looking at the evidence given by saying where the car was at time 2 minutes, we see that the average velocity during that time was 3 miles per minute. But, in this case, when we look at shorter intervals of time, we get a different story.

Now, let's look at the shorter interval of time that would happen if we looked at where the car was at the 1.1 minute mark. Well, the position at 1.1 is obtained by taking 1.1 and multiplying it by itself. We see that we're at the position 1.21 on our well-marked highway. The distance traveled between 1.21 and 1, subtracting, is that the car traveled .21 miles during the elapsed time of .1 minutes; dividing, we see that we get the car was traveling during that interval of time, that .1 interval of time, 1/10 of a minute interval of time, it was traveling at the average velocity of 2.1 miles per minute. You see, that's a different story than we had before. Before it said 3 miles per minute; now 2.1 miles per minute was the average velocity during that shorter interval of time. And we get a sense, by the way, that the shorter intervals of time are giving us a better indication of what the car was doing at the exact moment in which we're interested, because during that shorter interval of time, the car is not changing its velocity as much; doesn't have time to do so.

$$\text{Average speed} = \frac{p(2) - p(1)}{2 - 1} = \frac{4 - 1}{1} = 3$$

mi/min

Let's continue this laborious process by looking at the 1.01 moment of time. This is 1/100 of a minute after the 1-minute mark. Where

$$\text{Average speed} = \frac{p(1) - p(0.999)}{1 - 0.999} = \frac{1 - 0.998001}{0.001} = 1.999 \text{ mi/min}$$

was the car? Well, if we multiplied 1.01 by itself, we get 1.0201. That's the position on this well-marked road. We subtract 1 from it; we get .0201 is the elapsed distance traveled during the 1/100 of a minute that we're talking. It's very precise. And we see, now, that its velocity during that very short interval of time was 2.01 miles per minute. Now, this is getting a little interesting, that we're getting different answers. As we take different, shorter intervals of time we're getting different answers, but those different answers are tending toward one number, and this is going to be one of the main characteristics of calculus and of doing the derivative in particular. So, this is—we're going to do this yet one more time. And, in fact, we're going to do it infinitely many times, except, of course, we won't actuall y do it on

$$\text{Average speed} = \frac{p(1.01) - p(1)}{1.01 - 1} = \frac{1.0201 - 1}{.01} = 2.01 \text{ mi/mi}$$

camera infinitely many times, but we will do it infinitely many times.

Let's start with, once again, 1.001. So, this is 1/1000 of a minute after the time that we're interested in. We see how far the car went during that interval of time. It went .002001 miles, and it took 1/1000 of a minute to do so; when we divide, we get that the average velocity during that 1/1000 of a minute was 2.001 miles per minute. So, you can see that the values of these are—evidence that's taken by increasingly shorter intervals of time—are giving us a sense of the velocity of the car, getting closer and closer to 2 miles per minute.

Let's just do one more example, just to see it; namely, if we take the car slightly before the 1-minute mark, at .999; if we do the multiplication; do the division; we see that the indicated average

$$\text{Average speed} = \frac{p(1.001) - p(1)}{1.001 - 1} = \frac{1.002001 - 1}{.001} = 2.001/\text{mi/min}$$

velocity during that 1/1000 of a minute was 1.999 miles per minute.

So, what is the evidence telling us? The evidence is giving us a sense that the velocity of the car is converging; that all of this different evidence, each piece of evidence, taken from two instants of time, seeing how far the car went during that short interval of time, and then dividing by the elapsed time—all of that evidence, each one gives us a number, and those numbers are getting closer and closer to the single number 2. That process is called a limiting process. We're taking a limit of values that are getting closer and closer to 2, and then we say that at the instance of 1, that the value is 2. That's the concept of the derivative. The derivative—for every instant of time we have to do an infinite process; and when we do the infinite process, we get a single number, namely, the number to which all of that evidence converges.

Well, this is, as you can see, this is a very laborious kind of a process. Even for this car moving at this time where its position at every moment of time is given by the square of the time on our watch, we see that it took us a long time, and it was sort of boring, to figure out how fast the car was going at the 1-minute mark. But, we could do this at every single moment of time. That is, this long, laborious process could be undertaken at different points of time.

So, for example, this is a chart that accumulates all of the evidence that we had seen from our previous investigations. Namely, the evidence was that the average velocity during the time 1 and time 2 was 3 miles per minute; between 1 and 1.1 it was 2.1 miles per minute; between 1 and 1.01 it was 2.01 miles per minute; between .999 and 1 to was 1.999 miles per minutes; and we, therefore, conclude that the instantaneous velocity at time $t = 1$ is 2 miles per minute. So, this summarizes our previous information.

Now, let's consider this same question for different times. In other words, instead of thinking about the time when our watch says 1 minute, let's consider when our watch says something else, like .7 minutes. Well, if we look at .7 minutes, and we do this same laborious computation, that is, we take .7, that's our initial time, and then we look at a minute later and we take the average velocity, we get 2.4; and we take .7 and we just do 1/10 of a minute afterwards, we compute the average velocity, 1.5; we take 1/100 of a minute after .7, we get an average velocity of 1.41. So, in each case we're doing this characteristic process of the position, the final position of the car, the initial position of the car, dividing by the elapsed time. That's what each of these numbers represents.

Notice that as we get closer and closer to the .7, the time at which we're interested in the instantaneous velocity, when we take just 1/1000 of a minute afterwards, or just 1/1000 of a minute before, our computations are leading us to believe that the actual instantaneous velocity is the number 1.4. So, we conclude, by doing infinitely many experiments and increasingly shorter amounts of time, that the instantaneous velocity is 1.4 miles per minute. Well, that's the computation associated with the value .7.

I'm just going to quickly show you some other charts where we did exactly the same laborious process. In the case of—at the 1.4 initial time, we do this same computation, and we conclude that the numbers to which those average velocities are tending looks like it's tending towards 2.8; 2.8 is the instantaneous velocity at the time 1.4.

When we're looking at the initial time 2, once again we can do this laborious process of taking nearby times, and we see that those values are getting near 4 miles per minute. When we look at an initial time of 3, we see that our instantaneous velocity is converging down to the number 6 miles per minute. By the way, now we're sort of beyond the idea of a car moving, since that's 360 miles per hour, but it's 6 miles per minute.

The point is that each of these laborious computations gives us a value of the instantaneous velocity; but, we see a pattern. Look at our pattern here. We did each one of these values here, the instantaneous velocity in this column, at the time .7, we deduced that that instantaneous velocity, computed from an infinite process, was 1.4 miles per minute. At the time 1, the instantaneous velocity was 2. At the time 1.4, the instantaneous velocity was 2.8. At the time 2, the instantaneous velocity was 4. At the time 3, the instantaneous velocity was 6. We see the pattern. Do you see the pattern? The pattern is that the instantaneous velocity is always exactly twice the time. So, once we observe this pattern, we see that if the car is moving at a rate so that its position at the road is always measured by t^2, then its velocity is measured by $2 \times$ the time. And once we accumulate that into that very simple form, then we don't have to go through the laborious process at each moment of time to see what the instantaneous velocity is. We have one equation that tells us instantly the instantaneous velocity at each moment of time. And that's the power of the derivative.

So, the word of the day is derivative. And when you think of derivative, what I want for you to come to mind is this repetitious thing that we've been doing all during this half-hour. Namely, if we have a position of a car moving along a straight road, and we want to know if the instantaneous velocity at the time t, we compute its position at time $t + \Delta t$, where by tradition Δ means Δt is a small increment of time and viewed as an increasing small increment of time, but a specific small increment of time, we see where we were a little bit after or before t, we subtract where we were at time t to find the elapsed distance traveled during that very short interval of time, and then we divide by Δt—which is the elapsed time, the time between $t + \Delta t$ and the time t. The elapsed time was Δt. This characteristic fraction, just looking at this fraction, tells us the average velocity of the car between two instants of time. Notice there's nothing continuous about this: it's one instant, and then another instant, and we compute the—everybody can agree on what the average velocity is if you give just two instants because it's the net instants divided by the elapsed time, and then what we do is we do this process that's the limiting process as Δt becomes increasingly small and goes to 0. That is the process of the derivative.

> The derivative of a function $p(t)$ at time t is
> $$\frac{p(t + \Delta t) - p(t)}{\Delta t}$$
> as Δt becomes small.

So, we've now been introduced to one of the two fundamental ideas of calculus, the derivative. In the next lecture, we're going to take the same scenario of a car driving down the road, and introduce from it the idea of the integral. I look forward to seeing you then.

Lecture Three
Another Car, Another Crime—
The Second Idea of Calculus—The Integral

Scope:

The second idea of calculus helps us understand how to take information about tiny parts of a problem and combine this information to construct the whole answer. To develop this idea, we return to the scenario of a moving car. In preparation, we take our car out to a straight road, say to El Paso. We then videotape only the speedometer as the car moves. We show some friends the video of the speedometer and ask them to predict where we were at the end of an hour. The process by which they use information about velocity to compute the exact location of the car at the end of the hour is the second of the ideas of calculus—the *integral*.

Outline

I. In the last lecture, we introduced the derivative—the first of the two basic ideas of calculus.

    A. The derivative allowed us to settle one of Zeno's paradoxes of motion because it told us what we mean by the instantaneous velocity of a moving car or arrow.

    B. Note that we did not fall into the trap of trying to divide 0 by 0 to get the velocity at an instant.

II. The second fundamental idea of calculus arises from a scenario involving another car and another crime.

    A. In this scenario, you are kidnapped, tied up in the back of the car, and driven off on a straight road.

    B. You cannot see out of the car, but fortunately, you can see the speedometer, and you have a video camera to take time-stamped pictures of the speedometer. (There is no odometer in sight.)

    C. After 1 hour, you are dumped on the side of the road.

    D. How far have you gone?

III. What information can we extract from the videotape?

A. Let's take a simple case: The car was going at a constant velocity.

    1. At a constant velocity, computing how far we have gone in a given amount of time is easy. For example, if we go 1 mile per minute for 60 minutes, we will have gone 60 miles during that hour.

    2. If we go 2 miles per minute for 20 minutes, we will have traveled 2×20, or 40 miles. On a graph, this constant velocity would appear as a horizontal line.

B. Let's take a harder case: Suppose our velocity changes. How do we compute the distance traveled?

C. Let's look at examples where our velocity is steady for some time, then abruptly changes to another velocity, and so on. Suppose we travel at:

    1. 1 mile per minute between times 0 and 10 minutes,

    2. 2 miles per minute for the time between 10 and 20 minutes,

    3. 3 miles per minute for the time between 20 and 30 minutes,

    4. 4 miles per minute for the time between 30 and 40 minutes,

    5. 5 miles per minute for the time between 40 and 50 minutes, and

    6. 6 miles per minute for the time between 50 and 60 minutes.

On a graph, this changing velocity would appear as "stair steps" going up.

D. Our total distance traveled will be:
$$(1 \times 10) + (2 \times 10) + (3 \times 10) + (4 \times 10) + (5 \times 10) + (6 \times 10)$$
miles; that is, 210 miles. (We were speeding.)

E. We now know the process of computing the total distance traveled if we know the velocity at each time and the velocities are steady at one velocity for a while, then jump to another velocity and so on throughout the whole hour.

IV. We know, however, that velocities do not jump like this and, instead, increase smoothly from one velocity to another. Dealing with variable velocities involves doing a little at a time and adding them up.

A. Our strategy is first to underestimate the distance traveled, then to overestimate the distance traveled, then to determine that the distance traveled is somewhere between these two.

B. Let's consider a car that is moving at each time t at velocity $2t$ miles per minute. That is, at 1 minute, we are traveling at a speed of 2 miles per minute; at 2 minutes, we are traveling at 4 miles per minute; and so forth. Because our car is going so fast, let's just see how far it goes during the first 3 minutes of travel.

C. In order to discover the distance covered, we will compare our car traveling smoothly at an increasingly higher velocity to a second car that is moving in the "jerky" or "jumpy" fashion we have seen above. We will also break the 3 minutes into smaller intervals of time, for example, half-minute intervals.

D. We then add up approximations of the distances traveled by the second "jerky" car during each short interval to approximate the total distance traveled.

    1. During each short interval of one-half minute, the car changed velocities.

    2. But if we assumed that the second "jerky" car continued at a steady velocity equal to the initial lower velocity during each little interval of time, we would get an *approximation* of the total distance traveled during that interval (7.5 miles).

    3. Because our car is always speeding up, the distance traveled by the second, "jerky" car will be an underestimate of the total distance our car traveled.

    4. Similarly, if we assumed that the second car went at a steady velocity equal to the fastest velocity our car actually went during each little interval of time, we would get an approximation to the total distance traveled during that 3-minute interval (10.5 miles).

    5. This approximation to the total distance traveled would be an overestimate.

    6. The correct answer would have to be somewhere between those two estimates.

E. The smaller the intervals, the more accurate will be the approximation to the distance traveled. Let's try breaking the time into intervals $1/10^{th}$ of a minute long.

    1. Once again, we can get an underestimate (8.7 miles) and an overestimate (9.3 miles).

    2. Notice that these over- and underestimates are closer to each other than before.

F. As the intervals get smaller, it doesn't matter what velocity we select from the range of velocities of the car in the intervals, because the car doesn't change velocity much in the tiny intervals.

G. Using increasingly smaller intervals produces increasingly better approximations.

V. The exact distance traveled can't be found with any single division of the interval of time.

A. The exact answer is obtained by looking at infinitely many increasingly improved approximations.

B. The finer approximations get closer and closer to a single value—the limit of the approximations.

C. This infinite process is the second fundamental idea of calculus—the *integral*.

D. If we know the velocity of a car at every moment in a given interval of time, the integral tells us how far the car traveled during that interval.

E. Remember that the derivative was the limit of the average velocities as the intervals got smaller and smaller; likewise, the integral is the limit of the approximations as the intervals get smaller and smaller.

VI. We can use the same analysis to find out where we were at any moment.

A. In the example above where the speedometer always reads $2t$, where are we after 1 minute, 2 minutes, 2.5 minutes, 3 minutes? In each case, we will use this infinite procedure to see how far we traveled from time 0 to these times.

B. Let's look at the table and see if there is a pattern.

Instantaneous velocity: $v(t) = 2t$	
Time (min)	Distance
0.5	0.25
1	1
1.5	2.25
2	4
3	9
Position: $p(t) = t^2$	

We can make a shrewd guess as to where we were after any time t: Namely, it appears that we are at the t^2 mileage marker.

C. Distance traveled is, thus, the square of the time interval taken, or $p(t) = t^2$.

D. The integral can be used to find the position of a moving car at each moment, if we know the velocity at each moment.

E. The integral process involves dividing the interval of time into small increments, seeing how far the car would have traveled if it had gone at a steady velocity during each small interval of time, and then adding up those distances to approximate the total distance traveled. Therefore, the formula to determine the distance traveled between time a and time b is:

$(v(a) \times \Delta t) + (v(a+\Delta t) \times \Delta t) + (v(a+2\Delta t) \times \Delta t) + \ldots + (v(b - \Delta t) \times \Delta t$, as Δt becomes increasingly smaller.

F. By taking smaller and smaller subdivisions and taking a limit, we arrive at the actual value of the integral.

Readings:

Any standard calculus textbook, the section introducing the definite integral.

Questions to Consider:

1. Suppose the speedometer on your rocket ship reads exactly $3t^2$ miles per minute at each time t minutes. Use the integral process to compute how far you will have traveled after 1 minute, 2

minutes, 3 minutes. Do you see an expression in t that gives the same answer?

2. Suppose your velocity is always $2t$ miles per minute at each time t minutes. How fine must your divisions of the interval of time be between time 0 and time 3 in order to be certain that your summation in the integral process is definitely within 1 mile of the correct answer?

Lecture Three—Transcript
Another Car, Another Crime—
The Second Idea of Calculus—The Integral

Welcome back. In the last lecture we introduced one of the two fundamental concepts of calculus, namely, the derivative; and we saw that the derivative allowed us to actually settle one of Zeno's paradoxes in motion because it told us what we were to mean by the instantaneous velocity of a moving car or arrow. So, that was an accomplishment that we did, and notice that it never involved the mistake of dividing zero by zero. We always looked at points of time that were close to each other so that there was a difference between the increments of time so we were never dividing by zero.

In this lecture, we're going to introduce the second fundamental idea of calculus, that is, the *integral*. And, once again, it's going to involve an infinite process, and it's going to involve a process that comes about as a natural result of analyzing a situation once again of a vehicle moving along the road. And in order to keep in our theme of sticking to crime, we're going to involve this lecture with another crime. So, this is called "Another Car, Another Crime." So, this is what we're going to do.

So, suppose this time the crime is a much more heinous offense than just merely running a stop sign, as we had in the last lecture. In this case, the crime is kidnapping. So, here's what happened. You were an agent for the Mission Impossible crew, and you were kidnapped by some nefarious bad people and you were put in the back of a van. Now, you have to think of this as a van in this case. So, here you are in the back of a van, and you're put in the van, and you know where you were captured. You were put in this van. But, the van has these walls on the side of it, so you can't see out on this well-marked road; you can't see any of the markings on the side of the road. But, the car starts out and goes along, and there it is driving along, and the only thing that you can see inside this back of the van where you're thrust is through a little hole that looks into the front compartment of the van; you can see the speedometer of the van. So, you see the speedometer and that's all you see. And, by the way, you do not see the odometer. You don't see the one that tells you how far you've gone. All you see is the speedometer. And, fortunately, you have a video camera with you, and so you take this video film; you point this video camera at the speedometer and you watch this extremely entertaining vision of this speedometer going like this, you see, back and forth. So, you have this hour of film of this speedometer of this car.

Now, you're captured for an hour, and you know that you're going just on a straight road because it never turns; it's definitely a straight road. And at every moment you've got this video film of the speedometer; at the end of exactly one hour, you're thrown out the side of the road; the van drives away; and there you are at the side of the road.

Well, you've got to be rescued by your compatriots there at Mission Impossible headquarters, and you have your radio. You call them up and say, "I don't know where I am on this road, but I have the video of the speedometer." And so, you send it back electronically to Mission Impossible headquarters and they, then, look at this hour film of the velocity—namely, the speed—of the speedometer during that hour. Now, it's a pretty boring movie, but they're goal is to compute where you are on the road. Now, just think about it. You see, you have the information; knowing how fast you're going at every moment, you have the information that allows you to conclude how far you've gone on the road. And, as we figure out that strategy—the strategy by which you can take this moving picture of the speedometer and deduce from that the distance that you've traveled—that is the second idea of calculus, the integral.

Let's go ahead and do it. Now, what's the best strategy for thinking of, figuring out, difficult issues? The best strategy almost always, certainly in mathematics, is to take a very simple case. We do a simple case, and then we figure it out; we've taught ourselves something, and then we do more complicated cases until we can do more general cases.

So, let's begin with a very simple case. Here is the simple case: Suppose that the speedometer is the most boring possible movie because during the entire time of the hour in which you were sitting in the car, the speedometer never moved. It was exactly at one place the entire time. So, here's an example. Now, given this moving picture of the velocity; that is, the speedometer you see, is telling what was the velocity of that van at every moment of time, that's what the speedometer is telling us. Well, if the speedometer were just completely at the same speed the whole time, this extremely boring movie—sounds like an avant-garde thing, we should think about it. But, we could graph that velocity movie in a very simple way. Namely, at each moment of time we can just make a mark as to what the velocity was reading, that is, what the speedometer was saying. And, if it were completely a constant speed, we would just see this horizontal line, and that would be the velocity of this moving car. [The equation of the line would be $v(t) = c$.]

Now, let's suppose, for example—as before, we will be talking about miles per minute. So, let's suppose the velocity is 1 mile per minute; and it just sat there at the 1 mile per minute velocity and stayed at that one place during that whole hour. Well, how could we compute how far we went at the end of the hour, by the end of the hour?

That's very simple. We would say if we went 1 mile per minute for 60 minutes, then the total distance that we went is 60 miles. We multiply. We multiply the velocity at each time, if it's a constant velocity × the elapsed time, to get the distance that we traveled. So, if it were a constant velocity, we would be in very good shape.

Well, therefore, we've learned something. We've learned the very simplest case. But, we can do harder cases. Once we've taught ourselves simpler cases, we can do harder cases. For example, suppose that our car, the velocity, that is to say, that the speedometer that we were taking on our film, suppose that it didn't smoothly change. Suppose that it went at one speed for a certain amount of time, a certain number of minutes, and then instantly jumped to another speed for another number of minutes, and then instantly jumped to another speed for another number of minutes, and so on. So, let's be specific and look at a specific moving picture of the speedometer that we would want to analyze in order to rescue our compatriot here.

Well, here's an example. Suppose that during the first 10 minutes the speedometer read 1 mile per minute; and then it instantly jumped, during the next 10 minutes, to 2 miles per minute; and then during the next 10 minutes it moved to 3 miles per minute; and so on, 4 miles per minute, 5 miles per minute, 6 miles per minute. And, at the end the 60 minutes, we would have a graph of the velocity that was capturing what this motion picture of the speedometer showed us that would look like this. It's a step function. That is to say, it has a certain fixed value for an interval, and then another fixed value for an interval, and so on. How could we compute the distance that we traveled during that hour so that we know where to go and rescue our friend?

Well, it's easy to compute how far the van went during that first 10 minutes, because it went 1 mile per minute for 10 minutes. The product, velocity × time, is the distance traveled. So, we went 1 × 10; that's 10 miles. During the next 10 minutes we went 2 miles per minute for 10 minutes; so, the distance traveled is the product of the velocity, 2 miles per minute, × the time elapsed, 10 minutes—2 × 10 is 20 miles, and so on.

So, to compute the total distance traveled, we have to add up the distances traveled during each of those 10-minute intervals. So this is very simple, right? This is what we would do if we were tasked with the goal of finding how far that van went during the hour. Velocity × time + velocity × time + velocity × time, and so on, giving us a total of 210 miles during the hour. By the way, this car is really moving. At the last it's going 6 miles per minute, which is 360 miles an hour. But, these people don't mind breaking the law, so that's no problem.

What we have taught ourselves so far is how to compute the total distance traveled if the car had the property that the velocities went for an increment of time at a steady rate and then went at a steady rate and then went at a steady rate. But in real life cars don't do that; in real life, cars smoothly go from one velocity to another velocity. And so, if we looked at our actual picture of the speed of the car, and that is the speedometer, we might see it moving like this, gradually moving up and down or whatever it did. And, so, we've got to extend the strategy that we've so far developed to deal with the situation where the velocity is actually changing. So, let's go ahead and do this in another situation.

Suppose that in this moving picture of our speedometer, the velocity we're given to us by the time always staying at 2 × the time on our watch was the speed. In other words, it started at a stop; at time 0 it was stopped, which was probably good. And, then, as we went forward, at the 1-minute mark it was going 2 miles per minute; and then at the 2-minute mark it was going 4 miles per minute; at the 3-minute mark it was going 6 miles per minute; and so on. That would be a way that the speedometer may behave. How can we compute the distance traveled in this case?

Well, in this case we're not in a position to actually get the exact distance traveled during an interval of time; at least, it's not obvious how to do it, as it was when the velocity was constant. When the velocity was constant, it was clear what we should do. But, in this case, our velocity is varying. So, our strategy in this case is to do two things. One is, we're going to underestimate the distance traveled, and then we're going to overestimate the distance traveled, and then say that the distance actually traveled has to be somewhere between them. So, that's our general strategy. Well, how could we underestimate the velocity—I mean, the distance traveled?

So, one way to underestimate the distance traveled is to compute the distance that another car would have traveled had it always been going somewhat slower than the car that actually we're trying to compute. Well, since we know how to compute the distances traveled by cars going in this "jerky" fashion, going at a certain speed for a fixed time, then jumping up to another speed for another time. Let's consider the following: How about a car, this blue car is the car we're actually talking about, or the van, that's increasing in its speed entirely—at every moment it's increasing in time in speed. It's increasing in its velocity at every moment in time. Now, we could compare that to a car that stays still for a certain amount of time; and then, when this blue car has attained a certain velocity, then the red car begins going at that fixed velocity for an interval of time, and then it goes at the new velocity for an interval of time, and so on. So, the red car is going in this "jerky" fashion, whereas the blue car is continually increasing in its velocity. Let's go ahead and look and see graphically what this might look like.

Graphically, this is a graph of the velocity of the actual moving car. That is, we're assuming that it's velocity at every time t is $2t$. [The equation of the line is $v(t) = 2t$.] Well, these lower horizontal lines represent a movie of another related car whose velocity is always less than the given car. In the sense that, during the first interval of time—and, by the way, we're just going to think of the first 3 minutes of this motion instead of 60 minutes because we're going to think about dividing it up into a lot of pieces and 60 is too big a number. So, we're just going to think about how far this car went in the first 3 minutes of its travel. So our strategy for doing this is to say, let's divide our time interval into smaller intervals of time. In this case, we chose to divide the interval between 0 and 3 minutes into intervals of 1/2 minute. And, during the first 1/2 minute, we're imagining a car that just stays still for the first 1/2 minute; then it jumps up to go at 1 mile per minute for the next 1/2 minute. Now, notice that our car whose velocity is varying at the—prescribed by velocity at each time t is $2t$. At time 1/2 minute, the velocity of our moving car has attained 1 mile per minute. Now, our new car, the one that we're thinking about going slightly slower speed, then stays at that slower speed for 1/2 minute. Then it says, well what is the speed of our original car at time 1 minute? Well, it's going 2 miles per minute. So, our new car, the one that is moving in this "jerky" fashion, stays, then, at the speed of 2 miles per minute for the next 1/2 minute. And then it jumps up the speed that's been attained by the car at the 1-1/2 minute mark, which is 3 miles per minute, and stays at that speed for 1/2 minute, and so on.

Notice that the car that's moving in this "jerky" fashion is always going at a speed that is lower than the velocity of the given car. And, by the way, I have to catch myself, sometimes I'm saying "speed", sometimes I'm saying "velocity." Now, what's the difference? Velocity is speed with a direction. And sometimes we're going to later be talking about the car going around and turning backwards, in which case the velocity will be a negative number. And, so, I should be saying velocity each time, but as long as we're going forward it means the same thing.

Okay. So, here we have a case where the velocity is being approximated by a car that's going in this "jerky" fashion, but is always somewhat slower. Well, look, we know how to compute the distance that's been traveled by this car that's moving in a "jerky" fashion. That's what we taught ourselves before. So, how do we compute that? We say well, the car is going at 0 miles per minute for 1/2 minute; so, it hasn't gone anywhere during that first 1/2 minute. During the next 1/2 minute it goes at 1 mile per minute for 1/2 minute; that's 1/2 mile that it travels. During this next 1/2-minute interval, between time 1 and time 1.5, we're seeing that this car is going at 2 miles per minute for 1/2 minute. That's 1 mile. And so on throughout the intervals of time we see that it will go a total of 7.5 miles; it being the car that's going somewhat slower than the actual car whose velocity is always $2t$.

Well, that's fine, but we can do the same thing by overestimating, making an overestimate of how far the car goes by saying how about if we take a car that moves in the same kind of "jerky" way, but instead we consider it to be moving at a faster rate than the car actually moves during each sub-interval of time. So, in this case, we have a picture here, graphically, that we're considering a car that moves at the highest rate that the actual car attained during this first 1/2 minute, namely, 1 mile per minute during this first 1/2 minute, and then the highest rate that the actual car attained during the next 1/2 minute, namely, 2 miles per minute for the next 1/2 minute, and so on. That would be a car that went definitely further than the actual car during every single interval of time, and consequently, the totality would definitely be a larger number than the actual distance traveled by the car. If we do this computation of just the faster time × the interval + the faster time × the interval, and so on, we get 10.5 miles. That's how far this faster car would go during those three minutes.

Well, what can we conclude from this? We conclude that somehow the actual velocity of the car must lie between 7.5 miles, the distance traveled by the underestimate, and 10.5, the distance traveled by the overestimate. How can we get a better estimate?

Now, by the way, I hope that you now are sort of thinking back, and maybe with annoyance, that this same kind of repetitive strategy is going to happen that happened in the last lecture when we were talking about derivatives. How did we talk about derivatives? We said, well, if we take a smaller interval of time around the time of interest, we will get a different approximation that will be a better approximation of the actual instantaneous velocity. And now we're doing a similar thing. We're saying okay, we have a method for getting an overestimate of the distance traveled, and an underestimate of the distance traveled; we know the actual distance traveled is sandwiched between those two numbers.

How could we get a smaller sandwich? How could we get a smaller interval in which the actual distance has to appear? How? We simply take finer intervals of time. We divide our interval from time 0 to time 3 into tenths of a minute; tenths of a minute. Now, this is getting very laborious, but here's what we can do. We can take the interval of time between time 0 and time 3 and divide it into tenths of a minute. We have 30 intervals now. For each interval we can say we know what its initial time was and its final time was on our watch, so we know what the initial velocity of the car was and what the final velocity of the car was; it's going at $2 \times t$. So, if we want an underestimate of the distance traveled, we can assume that we have—we imagine we have a car that stays constantly at the lower speed for each of those 30 intervals of time. And, if we did that, we would have a long addition problem that involved 30 multiplications and additions, multiplying the velocity × the time + the velocity × the time + the velocity × the time, and this is capturing a picture of the car moving at little step values of velocity, but 30 of them this time.

Likewise, we can get an overestimate in the same way, dividing in these 30 intervals we can take a velocity that's slightly faster than the car actually moved during that 1/10 of a minute. And, if we do so, then during, for example, this first interval of time, the fastest—the first 1/10 of a minute, the fastest the car ever went during that 1/10 of a minute was .2 miles per minute; right at the end, that was its velocity. And so we assume, we say "What would have happened if a car had gone at that faster velocity during that first 1/10 of a minute? And the fastest velocity the actual car went during the second 1/10 of a minute?" And we added up those distances traveled, we would get a value of 9.3 miles. Therefore, we know that the actual distance the car must lie between our underestimate of 8.7 miles and our overestimate of 9.3 miles. So, you see, we're getting a narrower sandwich, a narrower window in which the actual distance must reside. Well, now you're getting the philosophy of the integral.

So, if we want to approximate the actual position of the car, the distance traveled by the car during that 3-minute interval, we could imagine dividing the interval of time into incredibly small intervals. And for each small interval, we could take the actual velocity of the car, the graph of its actual velocity, and imagine breaking the time interval between 0 and 3 minutes into very, very fine increments that we call Δt, and then for each one we could compute an overestimate by multiplying the velocity of an overestimate × the interval of time Δt—each width is Δt. So, we have the velocity overestimate × Δt, velocity overestimate × Δt, and so on. And, as we move along, you can see the highlight of where we are and the contribution to this summand of the distance traveled in each one as we add up this long, laborious addition process. We notice that the total distance, then, will actually come out to be sandwiched between an overestimate and an underestimate of this laborious addition process. [Total distance = $(v(\Delta t) \times \Delta t) + (v(2\Delta t) \times \Delta t) + (v(3\Delta t) \times \Delta t) + \ldots$ as Δt becomes small.]

As Δt becomes small, this addition process will become closer and closer to the actual value, you see, because at each moment of time if we have a step function that has just a little, tiny interval of time—like 1/100 of a minute—the actual car's velocity will not vary much at all between the step function velocity that's jerky but happening every 1/100 of a minute is not going to be much different from the actual velocity that changes slightly during those 1/100s of a minute intervals.

So, once again, what we're doing is we're devising a strategy for finding the totality of the distance traveled that is associated with a process of an addition problem, multiplication and addition, and then taking finer and finer increments of time gets us to the exact answer. Once, again, no one of our computations will give us the exact answer. Instead, the exact answer for how far we've traveled is a process of doing something infinitely many times. Namely, dividing the interval into small increments, multiplying how far you went during each of those small increments, if you stayed at a steady rate; and then taking the limit of those answers as we take finer and finer increments.

And, by the way, one of the concepts of calculus that is at the heart of both the derivative and the integral is this limiting process, and we'll see and talk about it a little bit in the next lecture, that this was one of the very big conceptual challenges for mathematicians for more than 100 years after calculus was invented, because this concept of coming down to one number that is approximated finer and finer by these approximations by taking smaller and smaller intervals, you know, what does it really mean? It's really quite a subtle kind of idea. But, we can see conceptually it makes a great deal of sense. If we wanted to figure out how far that person is one hour after he was kidnapped, the way we could do it is to divide our interval into tiny pieces and we'd see how far the van actually went.

Now, let's do some other kind of computation on this idea, and that is we were talking about the distance between time 0 and time 3, and we got the finest numbers actually computed told us that the actual distance traveled had to lie between 8.7 miles and 9.3 miles. If we had done that same laborious computation for 1/1000 of a minute, and then 1/10,000 of a minute, and so on, we would find that the sandwiching of those upper and lower estimates, over and under estimates, was getting closer and closer to 9 miles. And, so, in fact, the car went exactly 9 miles during that interval of time, between time 0 and time 3 minutes.

But, we might want to ask ourselves the question, can we think in a more—instead of always going to 3 minutes, suppose we looked at how far the van went between 0 minutes and, say, 1 minute? And how far did it go between 0 minutes and .5 minutes? And in each time we did this entire laborious process of dividing the interval up into little tiny pieces and adding them up, we would find that we have this same kind of a chart similar to the chart that we saw in the last lecture about derivatives; namely, there's a pattern here. If we look at the total distance traveled in the first 1/2 minute, we traveled .25 miles; that is, 1/4 of a mile. In the first minute, we traveled 1 mile; in the 1.5 minutes, we traveled 2.25 miles; in 2 minutes, 4 miles; in 3 minutes, 9 miles. If you notice the pattern, the pattern is that the distance traveled is just the square of the time elapsed between time 0 and the time that we're measuring. So, once again, once we see this pattern, we're in a position to instantly say how far that car went in any amount of car because we would just say between time 0 and time whatever it is, say, 4 minutes, you just square the 4 and that's the answer.

That is the strength of calculus; is that we're going to see that the formulas, for things like the velocity, give rise to a formula for the position of the car at each time. It's telling us something that could be computed by this very laborious process, and it's important to realize that the answer we get is the result of doing the laborious process whose result we can interpret as a meaningful thing; namely, how far the car traveled. That was a natural thing for us to do, and now we're seeing that it comes out to be a formula that makes it easy to compute that distance traveled.

So, the word of the day is the integral; and the integral is the, if you have a velocity function, then between any time a and a time b, we can compute the distance traveled by the car, the net distance traveled, by computing its velocity at time $a \times \Delta t$ + its velocity at just a slight moment later $\Delta t \times \Delta t + a + 2 \times \Delta t \times \Delta t$, and so on— this very long addition problem for any individual choice of Δt gives us an approximation of the distance traveled. [The integral of a function $v(t)$ between time a and time b is $v(a)\Delta t + (v(a+\Delta t) \times \Delta t) + (v(a+2\Delta t) \times \Delta t) + (v(a+3\Delta t) \times \Delta t) + \ldots (v(b) \times \Delta t)$ as Δt becomes very small.] If we take our Δt's increasingly small, that approximation gets finer and finer, and then, as we say, in the limit, we get this single answer. And that is the definition of the integral from a to b of this velocity function $v(t)$.

So, at this point then, we have introduced the two basic ideas of calculus: in the last lecture, the derivative, and in this lecture, the integral. Both of them associated with a car moving on a straight road. Because of the fact that we introduced them both in terms of a car moving on a straight road, in the next lecture we're showing the connection; we'll show the connection between the derivative and the integral, and see in what sense those are inverse processes of one another. So, in the next lecture I'll look forward to telling you about the Fundamental Theorem of Calculus. See you then.

Lecture Four
The Fundamental Theorem of Calculus

Scope:

The *Fundamental Theorem of Calculus* makes the connection between the two processes discussed in the previous two lectures, the derivative and the integral. Again, this theorem can be deduced by examining the generative scenario of the moving car. The derivative and the integral involve somewhat complicated procedures that appear unrelated if viewed in the abstract; however, they accomplish opposite goals—one goes from position to velocity, the other goes from velocity to position. The duality between the derivative and integral is exactly what the Fundamental Theorem of Calculus captures. This convergence of ideas underscores the power of abstraction, one of the global themes of this series of lectures.

Outline

I. The two fundamental ideas of calculus, namely, the methods for (1) finding velocity from position (the derivative) and (2) finding distance traveled from velocity (the integral), involve common themes.

   A. Both involve infinite processes.

   B. Both processes involve examining a situation with increasingly finer time intervals.

   C. Both processes involve deducing a single answer from the whole infinite collection of increasingly accurate approximations.

II. We looked at the same situation—a car moving on a straight road—from two points of view.

   A. Knowing where we were at every moment, we deduced our velocity at every moment—the derivative.

   B. Knowing our velocity at every moment and where we started, we deduced where we were at every moment—the integral.

   C. The two processes are two sides of the same coin.

D. Understanding implications of this relationship between these two processes is *the* fundamental insight of calculus. Indeed, it is known as the *Fundamental Theorem of Calculus*.

E. Specifically, suppose we are given a velocity function; that is, we are told how fast we are traveling at every instant between two times. We can find the distance traveled in a given time interval in two ways.

    1. First, we can compute the integral (by dividing the time interval into little pieces and adding up distance traveled over the little pieces).

    2. Second, if we know a position function whose derivative is the given velocity function, we can simply use the position function to tell us where we are at the end and where we were at the beginning, and then subtract the two locations to see how far we went.

F. Method 2 is a lot quicker. It does not involve an infinite number of approximations as the integral does.

G. These two ways of computing the distance traveled give the same answer. That's what makes the Fundamental Theorem of Calculus so insightful—it gives an alternative method for finding a value that would be hard or impossible or, at best, tedious to get, even with a computer.

III. The moving-car scenario presents a situation to analyze the Fundamental Theorem. Let's do so where the position function is $p(t) = t^2$, and the velocity function is $v(t) = 2t$.

A. We can find the distance traveled from time 1 to time 2 via the integral. Remember all the sums involved:

$(v(1) \times \Delta t) + (v(1+\Delta t) \times \Delta t) + (v(1+2\Delta t) \times \Delta t) + \ldots + (v(2 - \Delta t) \times \Delta t$, as Δt becomes increasingly smaller.

B. Then we can find the distance traveled knowing that $p(t) = t^2$ by subtraction. Because $p(2) = 4$ and $p(1) = 1$, the car traveled 3 miles.

C. We can consider other pairs of time values.

D. The summing process of the integral will always yield the same result as just subtracting the positions, because all these processes are referring to the same scenario of a moving car.

E. Suppose someone told us that the velocity of a car moving on a straight road at each moment t was $v(t) = 2t$ but didn't tell us the position function and asked for the distance traveled in the first 3 minutes. We could either add up pieces via the integral process or we could find the position function and subtract.

F. The fact that both processes yield the same answer is the importance of the Fundamental Theorem of Calculus.

G. The process of the integral (summing up pieces) tells us that the answer is what we want to know. That process refers directly to the commonsensical way of finding the distance traveled given the velocity.

H. If we can find a position function, it is much easier to just subtract.

IV. The fundamental insight relates the derivative and the integral.

    A. The process of finding instantaneous velocity from position is the inverse of the process of finding position from velocity.

    B. We have two ways that give the same answer to the question of how far we have gone given the velocity that we have been traveling.

        1. One is by the infinite process of adding up (the integral).

        2. One is by finding a function whose derivative is the velocity function (thus, it is a position function) and subtracting.

        3. The insight is that both processes give the same result.

    C. From an arithmetic point of view, the Fundamental Theorem notes that the process of subtraction and division that is at the heart of the derivative is the opposite of the process of multiplication and addition, which is at the heart of the integral.

    D. This insight has many other applications, which we will see in future lectures.

V. The development of calculus was an incremental process, as we saw when we spoke of mathematicians before Newton and Leibniz.

A. Newton and Leibniz systematized taking derivatives and integrals and showed the connections between them.

B. The development of calculus, however, was involved in considerable controversy.

    1. One type of controversy concerned who should get the credit for calculus, Newton or Leibniz.

    2. The second type of controversy concerned the validity of the ideas underlying calculus, particularly the tricky business involved in
taking limits. Let's take a few minutes to talk a bit about each of these controversies.

VI. Supporters of Newton and Leibniz had a lively and acrimonious controversy about who developed calculus first.

    A. Newton was quite averse to controversy, and this aversion made him reluctant to publish his work. Newton developed the ideas of calculus during the plague years of 1665–1666 when Cambridge was closed, but he did not publish those results for many years, in fact, not until 1704, 1711, and posthumously, in 1736. He did, however, circulate his ideas to friends and acquaintances in the 1660s.

    B. Leibniz was the first to publish his results on calculus. He conceived the ideas in 1674 and published them in 1684.

    C. In 1676, Newton learned that Leibniz had developed calculus-like ideas. Newton staked a claim on his priority in the invention of calculus by writing a letter to Leibniz.

        1. In this letter, Newton indicated his previous knowledge of calculus by writing an anagram.

        2. The anagram consisted of taking all the letters from the words of a Latin sentence, counting them, and putting all the letters in alphabetical order, as follows: "6a cc d æ 13e ff 7i 3l 9n 4o 4q rr 4s 9t 12v x."

        3. The sentence was, "*Data æquatione quotcunque fluentes quantitates involvente fluxiones invenire, et vice versa.*"

        4. This means, "Having any given equation involving never so many flowing quantities, to find the fluxions, and vice versa."

        5. Even the English version is not much of a hint of calculus.

D. Some British supporters of Newton felt that Leibniz got the idea of calculus from Newton's manuscript during a visit that Leibniz made to England in 1674. Newton's supporters hinted at foul play in 1699.

E. Modern historians believe that Newton and Leibniz independently developed their ideas.

F. In any case, the controversy led to a downhill trend in relationships between the supporters of the two men.

G. The controversy had a bad effect on British mathematics for a long time.

VII. The other controversy associated with calculus involved its validity.

A. One thing that we have to understand is that, at the time of Leibniz and Newton, ideas that we consider absolutely fundamental to even starting to think about calculus today simply did not exist at all, for example, the idea of function.

B. Most vague, though, was the idea of the limit. Neither Leibniz nor Newton had firm ideas of the limit.

C. The concept of the limit was not resolved until the mid-1800s.

Readings:

Any standard calculus textbook, section on the Fundamental Theorem of Calculus.

Boyer, Carl B. *The History of the Calculus and Its Conceptual Development.*

Questions to Consider:

1. The derivative of the position function $p(t) = t^4$ yields a velocity function $v(t) = 4t^3$. Given that fact, use the Fundamental Theorem of Calculus to compute the distance a moving object will have traveled between time 0 and time 2 if its velocity at each time t is $4t^3$. If you like, you can check your answer by using the definition of the integral to compute the distance traveled.

2. We are here at one moment and there at another time. Thus, we know how far we traveled. Now let's look at this scenario dynamically, namely, how did we get there? Explain how the Fundamental Theorem of Calculus shows the connection between the dynamic and static views of the world.

Lecture Four—Transcript
The Fundamental Theorem of Calculus

Welcome back to *Change and Motion: Calculus Made Clear.* We've, in the last two lectures, been introduced to the two fundamental ideas of calculus, the derivative and the integral. Those two fundamental ideas are for the purpose of finding the velocity from the position function; if you know where you are at every moment, how do you find your instantaneous velocity? That was the derivative. And, two, finding the distance that you traveled if you know the velocity that you're moving at every moment of time; that was the integral. And these two fundamental ideas involved some common themes. First of all, they both involved infinite processes. You can't take a derivative by just doing one thing and you can't take an integral by just doing one thing; they're both doing infinitely many things, dividing the time integral into finer and finer bits in order to get better approximations, and then taking a limit to get an actual one answer.

So, both processes also involve this concept of taking a scenario of a car moving down a straight road. They were introduced that way. And in fact, we were looking at this scenario of a car moving down a straight road, but we looked at it from two points of view. From the one point of view, we said if we know where the car is at every moment of time, we can deduce how fast it's going; what its velocity is at each moment. And on the other hand, in the last lecture, we saw that if we had a picture of the speedometer, that is to say, we knew the velocity at every moment of time, and we know where we started, then we could compute where it is that that car would end up. That was the process of the integral.

Well, realizing that both of these processes involved the same car moving down the same road tells us that these two processes are two sides of the same coin; that they somehow are both reflecting one common underlying reality, that car moving down the road. Well, when we understand the implications of that relationship between the two processes, that insight is the fundamental insight of calculus. In fact, it's called the Fundamental Theorem of Calculus. It's putting those two ideas together.

So, let's think, specifically. Suppose that we're given a velocity function; that is to say, we're thinking about a car that's moving on a straight road, and at every time we're told what the velocity is of that moving car. Well, we can find the distance traveled in two different ways, okay? Here are the two different ways we can find the distance traveled. First of all, we can use the method of the integral. The method of the integral is to take the velocity at every moment and dividing it up into little moments of time and seeing how far we traveled during each little interval of time, and adding them all up to get the total distance traveled. But, there's another way that we could get the distance traveled. Suppose that we knew the position function of this car moving on a straight road. In other words, at every moment we knew where it was now, and where it was then. Then our strategy of computing how far we went between one time and another time would be much more straightforward. All we would do is we'd say well, where did we end up? Where did we begin? Subtract.

Well, that second method—the where are we at the end; where are we at the beginning; and subtract—is a lot quicker because it doesn't involve that infinite process of dividing the time into little bits and doing these infinite number, I mean, many multiplications and then adding them together and then taking a limit. Those are two different ways of getting the same answer; namely, the distance traveled.

Well, now what we need to do is to say what the Fundamental Theorem of Calculus gives us. It gives us an insight for an alternative way of computing a value that would be difficult, or maybe even impossible, to compute—at best it would be tedious, even with a computer—to compute the distance traveled by that integral method; by taking little intervals of time and adding them together; that would be a long, laborious kind of process.

But, this moving car scenario, therefore, gives us a situation in which we can present the Fundamental Theorem of Calculus so that it is very clear. So, let's go ahead and see if we can actually understand it in a particular situation. Namely, suppose that we're in the situation of a car moving on the road where at every moment of time the position function is given by t^2. This was the scenario that we analyzed in Lecture Number Two. The speed was given at every time t, that is the position, not the speed, not the velocity, the position at every time t was given by t^2. We looked at our watch, we squared it to see the position of the car at that moment; and we saw that a way to compute the instantaneous velocity was to look at this characteristic procedure of taking a difference in position divided by the difference in time, and computing that as this Δt, the difference in time, became smaller and smaller. And, we discovered in taking, for example, the velocity—instantaneous velocity—at time 1; we did these computations; all of which told us that the instantaneous velocity of time 1 was 2 miles per minute.

We went on to compute the instantaneous velocity at many different times, and we saw that there was a relationship if the position function at every time was t^2, then the velocity function at each time was $2 \times t$. And this chart sort of summarizes some examples of values. Sometimes when we looked at our watch, if our position were t^2, we'd see that the speedometer would be reading $2 \times t$. So, here's t; here's the position; here's the velocity at each time.

So, the derivative was a process by which we took a position function and deduced the velocity function. The integral, on the other hand, was the inverse process. It took the velocity function and gave us a position function. Or, actually, it gave us the net distance traveled during a particular time. We'd have to know where we started to see what the position was.

So, here is a little chart that summarizes the values for the function $p(t) = t^2$, and the velocity, $v(t) = 2t$.

So, let's go about computing the distance traveled between two specific moments of time. The time $t = 1$ and the time $t = 3$, just to pick arbitrary numbers. Well, looking back at our chart it's very easy to see what—this is at time $t = 1$, the position is 1 on the road. We look out the window it says we're at mile marker 1. At time $t = 3$, we look out the window and it says we're at mile marker 9. So, it's very easy to compute the distance traveled between time 1 and time 3. Namely, it's $9 - 1 = 8$. So, the distance traveled is 8.

On the other hand, the distance traveled could be computed in an alternative way. The distance traveled could be computed by doing this process of taking the infinite—infinitely dividing the interval of time between time 1 and time 3—and looking at this characteristic sum of the speed $\times$ the time interval traveled, adding it together to the next approximation, the next approximation, so that we can use this definition of the definite integral to get the total distance traveled. So, the Fundamental Theorem of Calculus tells us that if we have a position function, it entails the existence of a velocity function.

That's what the derivative does. If we know the position function that generates the velocity function, then we know that the integral process, all that multiplying and adding process that gives us the integral that will definitely tell us the distance traveled between the time we start and the time we end up, we can see that that would also be given to us by taking the position function that generated that velocity function and just plugging in the final value minus the beginning value to find the net distance traveled.

Okay. Let me see if we can do another example here. Suppose that we are traveling with a velocity function given by $v(t) = 3t^2$. Now, what that means is that every time we look at our watch and we look at our speedometer and we see that the velocity given by that speedometer, or velocitometer, is given by three times the time told on our watch squared. Now, suppose that we set ourselves the challenge of saying how far did we travel between time $t = 1$ and $t = 4$? Well, there are two methods to do it. The two methods to do it would be 1) to use the definition of the integral, the integral process, because, remember, that was the natural way to compute the distance traveled. We broke the interval up between 1 and 4 into small intervals of time; we approximated how far we went during each of those small intervals of time, and added them up. And then we took finer and finer intervals of time and took a limit to get one single answer. That was our strategy for figuring out the distance traveled. So, that definitely gives us the distance traveled.

On the other hand, another way we could find that same answer would be to say can we find a position function so that the derivative of the position function—that is, the derivative, remember, is giving us the velocity at each time. Well, if we could find a position function whose derivative is the velocity function with which we were first trying to deal, then we could use this much simpler way of finding the distance traveled by just saying, where does the position function tell us we are at the end, at time $t = 4$, and where does it tell us we were at time $t = 1$, and just subtract.

So, that is the beauty of the Fundamental Theorem of Calculus. It lets us avoid this infinite addition kind of process—it's infinite because we have to use smaller and smaller intervals—and, instead, if we can find a function, in this case the function is $p(t) = t^3$, with the property that the instantaneous velocity at every moment for a car that's at position t^3, its instantaneous velocity at every moment is $3t^2$.

Knowing that, we know that the total distance traveled is the position at time 4 minus the position at time 1, and that's the answer. The position at time 4 is 4^3— $4 \times 4 \times 4$, that's 64, $- p(1)$ is 1^3—that's 1—so the total distance traveled is 63. And, if we were to do this long, laborious addition process of the integral, we would get 63. So, that is the insight of the Fundamental Theorem of Calculus. It shows that the derivative and the integral are inverse processes to each other.

So, notice that what we've seen here is that the infinite process of adding things up, which is the integral, gives us this distance traveled; and the other was this derivative process, that is, finding the velocity function from the position function, is a question of subtracting and then dividing. So, the insight is that both these processes give the same result. That's what I've just been saying. But, think of it from an arithmetic point of view. So, the Fundamental Theorem notes that the process of subtraction then division that's the heart of the derivative gives us the opposite thing to the process of multiplication and addition, which is the integral. So, this insight has many, many applications because what it allows us to do is to take this integral process, which is laborious but it tells us what we want to know—in this case how far the car went—and we'll see that it tells us all sorts of other things that we want to know, but that we can actually compute the answer by finding a function whose derivative is the thing that we want to add up in that laborious integral kind of way.

Well, at this point, I thought it might be a good break for us, now that we've seen the three fundamental ideas of calculus. We've seen the derivative and the integral, the two fundamental ideas of calculus are those two; and, then, the connection between them is the Fundamental Theorem of Calculus. So, we can sort of take breather here and declare a victory over these really wonderful ideas. And, in all of the future lectures, we are going to see how rich these ideas really are because they apply to so many things that can be interpreted in so many different ways. But that's for the future lectures. I thought right now would be a good break for us to talk a little bit about the history.

We've already shown that the history of the calculus, of the development of the calculus, was definitely an incremental process. There were ancient roots to it; Eudoxus and Archimedes both used processes that were very reminiscent of the integral. Then Fermat and Isaac Barrow and many others developed ideas that were close to the derivative. And it was Newton and Leibniz who actually systematized the taking of derivatives and integrals; and they were the ones who showed the connection and pointed out the connection that we just saw in the Fundamental Theorem of Calculus.

This idea of calculus and the Fundamental Theorem is really a wonderful accomplishment, and in fact, it's been celebrated in many different ways. One of them is a couplet from a poem by Alexander Pope, which is I think really a wonderful tribute to Newton. It says the following: "Nature and nature's laws lay hid in night; God said, 'Let Newton be,' and all was light." So, there's a tribute to Newton. I wanted to say what Leibniz had to say about Newton. Leibniz was the other co-inventor of calculus, and he had a comment to say about Newton. I want to tell you that I've been on many committees in the mathematics department to hire people or to promote people, and you get these letters of recommendation, of course, about people. Letters of recommendation often are very glowing. They say "Oh, this is the best—one of the three best people in the world." In fact, there are many people who are among the three best people in the world, it turns out, even though they're math… But, here is a letter of recommendation that you don't read every day, and if I were on a hiring committee and I read this one, it would certainly get their attention. So, this is what Leibniz had to say about Newton. He said, "Taking mathematics from the beginning of the world to the time when Newton lived, what he has done is much the better part." So, there's an amazingly comprehensive assessment of Newton's contributions.

But, we don't have to go back to the time of Newton and Leibniz to see their—how important these contributions are. There's a book called, *The Hundred Most Influential Persons in History*. It's sort of amusing. This author just decided to write down what he thought were the hundred people who contributed the most in history. And on that list, number two is Isaac Newton, because of the calculus and, of course, his laws of physics, many of which were related.

But, the development of the calculus actually received considerable controversy. It was involved in a lot of controversy, and the controversies were really of two types. One type was the controversy about who should get the credit for calculus; should it be Newton or should it be Leibniz? And the second kind of controversy about calculus concerned the validity of the ideas that underlie calculus itself; and, in particular, the tricky business that's involved in this taking of limits. That was a very big conceptual obstacle for people. And so, what I'd like to do is just take a few minutes and talk about each of these two controversies.

We'll begin with the personal one about credit. It turns out that Newton was, apparently, sort of pathologically averse to controversy, ironically. And in fact, it's partly because of his aversion to controversy that he was embroiled in possibly the biggest controversy concerning priority—credit—for a discovery of any controversy in the history of mathematics and, maybe, science. And, in a sense, it was cause and effect because his aversion to controversy made him extremely reluctant to publish things. So, he would come up with ideas—and he would write them down, by the way—but he wouldn't publish them; he wouldn't make them public. So, then, the question of who actually came up with ideas first was sort of problematical, as we'll begin to see.

Newton actually developed the basic concepts of calculus during the middle of the 1660s. He was a student at Cambridge University, and it closed for two years, in 1665 and '66, owing to the plague. So, during that time Newton went to his aunt's farm and spent these two years thinking. During that time he developed these seminal ideas, his seminal ideas not only of calculus, but also of physics. Of course, as a professor it does make me think, "How well would our students do if we simply close the school and let them go back and think for themselves?" But, that's another story. But, in 1669, Newton actually wrote a paper on calculus, but it wasn't published; he didn't publish the paper, he just wrote it. In 1671, he wrote another paper on calculus; didn't publish it. He wrote another paper in 1676 and didn't publish it. In fact, all three of these papers were eventually published. The one he wrote in 1669 was published in 1711; that's 42 years after he wrote it. The one he wrote in 1671 was published in 1736, and that's nine years after he died. And the paper he wrote in 1676 was published in 1704; it was actually a part of the appendix to his *Opticks*.

But he did talk about the calculus somewhat in the *Principia*, which was written in 1687, but the arguments that he gave in the *Principia*, for his physical arguments, he basically discovered them using the methods of calculus, and then translated them into the much more laborious and old-fashioned methods of mathematics that were around before calculus, because calculus was not, at that time, a normally accepted method. So, none of his works on calculus were published when he developed those ideas. But, we know that he did develop those ideas because he wrote them and circulated those ideas to his friends and acquaintances. So, he definitely had developed the ideas of calculus in the 1660s; and, he used the techniques of calculus in all of his scientific work, and they appear in the *Principia*.

Well meanwhile, Gottfried Wilhelm von Leibniz independently invented calculus. And, his invention of calculus, he claims to have been sometime in the middle of the 1670s, so people think that it was probably somewhere in the neighborhood of 1674 that he got the ideas of calculus, and he published them in 1684; that's 10 years after he got the ideas. But, notice that 1684 is three years before Newton's first account of calculus appeared at all; his first account being in the *Principia*. So, I wanted to read you the title of Leibniz's paper on it and a little commentary on about how well it was received. Leibniz wrote his memoir on calculus; it was six pages long and it appeared in *Acta Euroditorum* of 1684. The title of his paper is this: "A New Method for Maxima and Minima, As Well As Tangents, Which Is Not Obstructed by Fractional or Irrational Quantities." One of the things that both Newton and Leibniz did were to generalize methods that generally had existed about derivatives and so on to involve more general classes of functions, and so, this was what he advertised in this paper. But, I thought it would be amusing for you to hear how it was received. The Bernoulli brothers were—I mentioned that there was a family of eight Bernoullis who did a lot in developing calculus, particularly on the Continent, and one of the Bernoulli brothers had this to say about Leibniz's paper of 1684. He said it was—it being the paper—was "an enigma rather than an explication." Apparently, the paper was extremely difficult to make any sense of. For one thing, it contained many misprints; and, in the other, it used this strategy of exposition that is, unfortunately, still too common in mathematical circles of trying to be extremely terse and consequently not explaining how the ideas came about.

But, then, we come to the question of the controversy about the priority, about who came up with these ideas. When Newton began to realize that Leibniz had the basic ideas of calculus—and Newton began to be aware of this in the 1670s, in the middle of the 1670s—Newton's response, he wanted to make sure that he got credit for calculus. So, he wrote a letter—and he actually wrote it—it eventually went to Leibniz, but it went to Oldenburg first in 1676, and Newton wanted to indicate his previous knowledge of calculus. The way he did it was he took a Latin sentence, and then he scrambled the letters; in other words, he made an anagram of the letters in this one Latin sentence. So the anagram consisted of removing all the letters and just putting them in order. So, here is what Newton wrote. He said there were six As, two Cs, a D, 13 Es, two Fs, seven Is, and so on; and these were the letters of a sentence. Now, the sentence that he wrote was the following. It says "*Data æquatione quotcunque fluentes quantitates involvente fluxiones invenire, et vice versa,*" which means, "Having any given equation involving never so many flowing quantities, to find the fluxions, and vice versa."

Now, even to a mathematician, this means very little. This sentence—it encapsulated Newton's thinking about derivatives, but it is a little bit obscure. And, when you read it, it doesn't mean that much. But, it does capture his concept of the idea of the inverse property of the derivative and the integral, fluxions was his word for the derivative. So, he tried to establish his priority in that fashion. But, then, later, some proponents of Newton made accusations that Leibniz had actually read some of the manuscripts of Newton's before he got his ideas.

Well, modern historians believe that the two inventors were independent—Leibniz and Newton independently thought of calculus, and you can see that it is incremental and it came about close to other work. But, Leibniz had published first, so people who sided with Leibniz said that Newton had stolen the ideas from Leibniz, and it became just a huge mess. It embroiled the British mathematics in opposition to mathematics in the Continent, and the two marked camps didn't talk to each other much and it really impeded the development of calculus in Britain compared to the Continent. So, it really was a sort of sad thing.

Well, there was another controversy associated with calculus, and that had to do with the validity of the reasoning. Now, you have to understand that the way we understand calculus at this time is, you know, fairly clear, and the things that we think of as absolutely fundamental to even starting to think about calculus didn't exist for them. They didn't have a clear idea of function. We're going to be talking about functions and graphs of functions; that wasn't something that they had a clear idea of until the 1690s, until after calculus was invented. But far more vague than that was the idea of limit. Remember, the idea of limit for both the integral and the derivative was taking these smaller and smaller things, getting approximating numbers, and then seeing that those numbers came to one value. The problem was that neither Leibniz nor Newton could possibly have—and did not have—a clear idea of this concept of limits. I wanted to read you a quote about Newton, talking about limits that might show you the clarity with which he mentioned them. So, here's Newton on limits. He says—he was asked about the meanings of the terms evanescent quantities and prime and ultimate ratios. Evanescent quantities—these are sort of disappearing quantities. And, he said the following:

> But the answer is easy, for by the ultimate velocity is meant that with which the body is moved neither before it arrives at its last place, when the motion ceases, nor after, but at the very instant when it arrives. And, in like manner, by the ultimate ratio of evanescent quantities, is to be understood the ratio of the quantities not before they vanish, nor after, but that with which they vanish.

You see, this is the idea of the limit, and he didn't have this very clearly in mind.

Well, people derided Newton and Leibniz for talking about infinitesimals, which they did, little, tiny infinitesimally small amounts of time, and about—there was a phrase that was used to ride them, calling Newton's quantities—evanescent quantities—of the ghosts of departing quantities. You see, these things were not at all clear, you see?

There was a particular opponent to calculus by the name of George Berkeley, who wrote an attack on calculus called *The Analyst,* and I just wanted to read you the title of this just because it's fun. Here's what it says. The title is "The Analyst," but the subtitle is the fun part. It says, "The Analyst; or A Discourse Addressed to an Infidel Mathematician."—and by the way, this referred to Newton's friend, Edmund Halley, not to Newton himself, who was a very devout religious person—"Wherein it is Examined Whether the Object

Principles and Influences of the Modern Analysis Are More Distinctly Conceived or More Evidently Deduced Than Religious Mysteries and Points of Faith. 'First Cast the Beam Out of Thine Eye, and Then Shalt Thou See Clearly to Cast Out the Moat Out of Thy Brother's Eye.'" That was the subtitle; apparently longer titles than occur in this day and age.

In any case, the concept of the limit wasn't actually resolved until the middle of the 19th century, in the 1850s. So, it really was a great challenge. People didn't have a good idea of what the real numbers were, and they didn't have a good idea of the limit process until that much later period. So, there really was a substantial issue involved in understanding the basis of calculus.

Well, in the next series of lectures we'll return to the mathematics and talk about the derivative and see it in various applications, particularly graphically, and then algebraically. I look forward to seeing you then.

Lecture Five
Visualizing the Derivative—Slopes

Scope:

Motion and change underlie our appreciation of the world, both physically and in many other realms. Change is so fundamental to our vision of the world that we view it as the driving force in our understanding of most anything. Frequently, the dependency of one variable on another is most easily described visually by a graph, for example, a graph that shows position as a function of time. The concept of the derivative provides a method for analyzing change. We explore the relationship between the graph of a function and its derivative. For example, we observe that an upward-sloping graph signals a positive derivative. Superimposing the graph of the derivative on the graph of a function reveals a visual relationship between a function and its rate of change.

Outline

I. In this lecture, we will look at the derivative and its relation to graphs.

    A. Graphs show a relationship between two dependent quantities.

    B. For example, when referring to our moving car, our interest in the derivative is to try to understand how the change in time affects the change in the position of the car.

II. Change through time is of fundamental interest in many settings.

    A. Physical motion is change in position over time.

        1. Understanding such change over time is important.

        2. Cars moving, the idea of velocity, is a basic example.

    B. In biology, we can consider the change in human population in the world from 1900–2000, for example.

    C. In economics, we consider changes over time to prices, employment, production, consumption, and many other varying quantities. With a graph, we can show the changes in the Dow Jones Industrial Average over the last century.

D. Understanding many important issues involves analyzing change in a characteristic over time. A graph can display the power output of the Chernobyl nuclear power plant on April 25–26, 1986. The steepness of the curve at 1:23 a.m. tells us how quickly the power output changed in a short amount of time, leading to the Chernobyl disaster.

E. These changes can be visualized using graphs.

III. Let's again analyze the velocity of a car given its position on a straight road.

   A. Let's look at the position graph and see how the velocity is related to the graph.

      1. First, looking at any position, we can figure out how fast the car is going.

      2. From the graph, we can deduce features about the motion of the car.

      3. If the graph is going up, the car is moving forward. If the graph is going down, the car is moving backward.

      4. The top or bottom of a graph means the car is momentarily stopped. Its velocity is 0.

      5. A steep graph corresponds to high velocity.

      6. A straight line means constant velocity—horizontal ↔ stopped; 45° upward to the right ↔ velocity of 1 mile per minute forward; 45° down to the right ↔ velocity of 1 mile per minute backward.

   B. Let's be more quantitative in our description.

      1. The steepness of the graph corresponds to the velocity. How exactly does it correspond?

      2. A straight-line position graph corresponds to a velocity equal to the slope of the line, where the slope is a quantitative measure of the steepness.

      3. Slope of a line is just the ratio of upward motion over sidewise motion, or the vertical change divided by the horizontal change.

         a. A straight line going upward to the right has a positive slope.

         b. A straight line going downward to the right has a negative slope.

         c. A horizontal line has slope 0.

IV. Consider now motion with varying velocities.

A. Recall the idea of the derivative and how that process gave the velocity. Let's see it with a curved graph of position for a car with varying velocities.

B. We look at nearby values and draw a straight line. The slope of that line is the change in position divided by the change in time.

C. As Δt becomes increasingly smaller, those straight lines converge to a tangent line whose steepness is telling us the instantaneous velocity.

D. Let's magnify the graph. Magnified, the graph looks more like a straight line, and the tangent line and the graph appear to coincide.

E. In general, the derivative at a point gives the slope of the line we would see if we magnified the graph.

F. Derivative, then, gives two equal quantities: (1) the velocity of the car and (2) the slope of the tangent line.

G. An important concept to remember is that smoothly curving lines, when viewed very close, look like straight lines. That is why the Earth looks flat to us even though we know it is curved.

V. We can see acceleration in the graph of a moving car.

A. Looking at examples of graphs, we can see where the velocity is increasing and decreasing.

B. We see that an upward-cupping graph corresponds to an accelerating car and a downward-cupping graph corresponds to a decelerating car.

C. Acceleration measures the change in the velocity over time.

D. Acceleration is itself a derivative—the derivative of velocity, or equivalently, the second derivative of position.

VI. In general, we observe the following relationships between functions and their derivatives:

A. A function is increasing if and only if its derivative is positive.

B. A function is decreasing if and only if its derivative is negative.

C. A function is "flat" if and only if its derivative is zero.

D. A function is concave up if its derivative is increasing, or equivalently, if its second derivative is positive. (This is the observation about acceleration we made earlier.)

E. A function is concave down if its derivative is decreasing, or equivalently, if its second derivative is negative. (This is the observation about deceleration we made earlier.)

F. Given a graph of a derivative, we can sketch the graph of the function, and likewise, given a graph of a function, we can sketch the graph of the derivative.

G. By sliding a tangent line along a graph and recording its slope at each point, we can generate the derivative graph.

VII. Let's look at the whole trip.

A. If you take a trip, you can easily compute what your average velocity was by taking the total distance between the starting and ending points and dividing by the time it took to cover that distance.

B. On a graph, that process is figuring out the slope of the line between the beginning and ending points.

C. Although your velocity may have varied during the trip, at some point, your instantaneous velocity will be exactly equal to your average velocity.

D. This reasonable observation is known as the *mean value theorem* (where *mean* means "average" rather than "cruel").

VIII. Let's summarize the relationship between a graph and the derivative.

A. The derivative of a graph at any point is equal to the slope of the tangent line.

B. If we magnify a smoothly curving graph, it will look like a straight line—the tangent line.

C. The perspective of a smooth curve looking like a straight line allows us to deduce various implications. One is known as *L'Hôpital's Rule*, after the man who wrote the first calculus textbook.

1. L'Hôpital's Rule states that the limit of a ratio of two smooth functions, both of which approach 0, is equal to the limit of the ratio of their derivatives.
2. A historical feature of L'Hôpital's Rule is that L'Hôpital did not discover it. He bought it from one of the Bernoulli brothers.

IX. When Newton and Leibniz defined the derivative in the 17th century, they used different words and different notation.

    A. Newton used a dot over a varying quantity to stand for the derivative. The problem with that notation is that an errant fly was capable of taking derivatives if it left its mark in the wrong place.

    B. Most people now use the notation for derivative that Leibniz introduced. Leibniz's notation includes the fundamental feature of the derivative as a quotient of changes, with Δ's becoming d's.

        1. For example, if $p(t) = t^2$, then

$$\frac{dp}{dt} = \frac{d}{dt}(t^2) = 2t.$$

        2. Often, y is a function of x, and the notation for derivative is $\frac{dy}{dx}$.

        3. This notation reminds us that the derivative arises from looking at ratios.

        4. The letter d reminds us of "difference," suggesting that the top value is, for example, a difference in the position of the moving car at two times, while the bottom is the difference in the time.

    C. If the function is presented as $p(t)$, then another notation for the derivative is $p'(t)$.

Readings:

Any standard calculus textbook, section introducing derivatives as slopes of tangent lines and section describing the connection between the graphs of functions and the graphs of their derivatives.

Questions to Consider:

1. When we look at a circle, we see a curve. Why is it that when we magnify a circle a great deal, it no longer looks curved?

2. Understand slopes of lines. That is, how does the slope measure the steepness of a line? Why is the slope of a line the same at each point of a line? Is the angle from the horizontal of a line doubled if the slope is doubled?

3. Draw a graph of a function—any function—and call it $p(t)$. Where is it increasing, decreasing, and constant? Where is it concave up or concave down? Can you sketch its derivative, $p'(t)$? Suppose what you drew first is actually $p'(t)$, can you sketch $p(t)$?

Lecture Five—Transcript
Visualizing the Derivative—Slopes

Welcome back to *Change and Motion: Calculus Made Clear*. In the previous lectures we've introduced the basic concepts of calculus—the derivative and the integral—and talked about how they're related to each other through the Fundamental Theorem of Calculus. But today, what we're going to do is begin a series of three lectures about the derivative as it relates to and reflects its presence in different manifestations. Today we'll talk about it's relation to graphs, graphical relationship; in the next lecture we'll talk about it's algebraic—the algebraic way of looking at derivatives; and, then, in the next lecture we'll talk about derivatives as they apply to things, other things in the world. The strength of the derivative and of the integral is that all of them can be viewed in these different scenarios, and in each case we see a richness and a relationship of this concept—of in this case, derivative—with, in today's lecture, the graphs.

So, let's first of all take a moment to think about how—what we're talking about when we talk about a graph. What is a graph? There are many instances in the world in which we are trying to relate two dependant quantities. In the case that we were talking about of a car moving down a straight road, we said at every time, the car is in a given position. So, that is an example of a function, because at every time you have a position.

Well the interest of the derivative is to try to understand what the change is; it's measuring how quickly a change in the time affects a change in the position, how quickly does the change in the position—what's the instantaneous velocity? Change in position with respect to change in time. And understanding change with respect to time is a very important concept that comes up in many other settings besides simply a car moving down a road; some settings that are not physical settings. For example, if we talk about the population of the world over time, we can draw a graph that captures this population change of the world over time, and we notice that in this graph of the population of the world in the century 1900 to 2000 that we actually have a steeply increasing curve. It's not a straight line; it's a curve that population slowly grows at the beginning, and then the rate at which it increases moves sharply upward. The population increases as a fast rate near the year 2000, and by looking at this graph, we can get a sense of how the population is changing with respect to time.

Here's another example of a graph. This is the Dow Jones Industrial Average over the last century; and it, too, has this property of having places at which it is increasing more slowly or more quickly.

Here's a very interesting graph. This is a graph that measures the power output at a nuclear power plant. At this nuclear power plant, this is at a particular day, August 25, 1986, and on this particular day there was the normal operation level of the power plant, how much output there was. It started fine; and then, at 1 a.m. it declined. So it was producing less power than ordinarily was the case. At 2 p.m. of that day it leveled off; and at 11:10 p.m. it made a sharp descent of less and less power being generated. Suddenly, at 1:23 a.m., the graph tells the story. Suddenly there were huge amounts of power being put out by the power plant; and this is, in fact, this particular day, and it's a particular place; this is the graph of the Chernobyl Power Plant on the day that it went critical. So, sometimes graphs tell us very compelling stories, particularly looking at how the steepness of the curve, which is telling us how quickly things are changing; in this case, how quickly the power output is changing, in just a short amount of time.

So, we're going to be discussing the concept of looking at a graph, and then trying to see how the derivative is associated with the graph. Let's take our example, again, a specific example here, of a car moving in a straight road—and a graph that captures the information of that car moving on a road. So, once again, let's imagine that our time axis is this horizontal axis, and at each moment of time we record where the car is on this straight road by making a mark and creating, thereby, a graph. So, if we wanted to record the motion of the car on this straight road, and if we thought of the straight road as the y-axis here—the vertical axis—we could record the motion of the car in the following way: We could move a dot along this curved graph, and the dot is moving so that its horizontal speed is constant, because we're thinking of the horizontal axis as telling us the way time is proceeding. And as it's proceeding, the car is moving on this straight road; it goes up to the 5-mile marker at time 1 minute, and then it descends back to the 1-mile marker at time 3 minutes, and then it descends up to the 5-mile marker at time 4 minutes ... and you can see that the car is just moving on this vertical, straight road, but as we move the dot along the graph, it's displaying the fact of the car's motion on the straight road. So, that's what the graph is telling us.

Let's look at the graph and try to interpret aspects of the graph with respect to the motion of the car. Well, the first thing to notice is that if the graph is going upward; that is, as we move to the right, if the graph is increasing, that means that the car is moving forward on the road. It's moving vertically. If the car, if the graph, is moving downward, that corresponds to the car's motion backward on the road. That corresponds to a negative velocity; meaning that we're thinking of negative being downward, positive being forward. And, at a place like this at the top, that's a place where the car is stopped momentarily as it changes direction from going forward to going backward. Notice that if the car has moved quite a long distance in a short amount of time—like going from the 1-mile mark to the 5-mile mark in 1 minute—that corresponds to a rather quick speed; whereas, if the graph is more horizontal—where it's not moving very far in a period of time, like between here and here—those correspond to slower speeds. It's now our—the next goal of the lecture is going to be to pin this down. Can we be absolutely precise about how quickly the car is moving by looking at the graph?

Well, let's look at some examples of graphs of cars moving on a straight road, but in this case some simpler graphs. Suppose that we have a graph of a car that just looks like a diagonal line, a 45-degree line. Well, this is an example of a car that is moving at exactly 1 mile per minute because it proceeds 1 mile along the road for every 1 minute of elapsed time. And so, the graph that is a straight diagonal line of 45-degrees corresponds to a velocity of 1 mile per minute, and a constant speed of 1 mile per minute.

Let's look at this example of a car moving on a straight road. Here, the graph has a steeper slope to it. In fact, for every 1 unit of horizontal distance, this straight line changes by 2 units of vertical change. That corresponds to a car where every minute corresponds to a 2-mile change, which is a speed of 2 miles per minute. Now, notice, that it's the steepness of the line that is telling us the velocity of the car, and the steepness of the line is measured by taking the— of a straight line—is measured by taking how much vertical distance is accomplished divided by how much horizontal time it took to accomplish that. It's a ratio of the vertical change divided by the horizontal change. And that ratio has a name. It's called the slope of a line. Notice that a straight line, if it's going upward, has a positive slope and, in this graph, you see a line going diagonally down to the right, and that has a negative slope because for a positive 1-minute increment, there is a negative change; and therefore, the ratio of rise over run is a negative number divided by the elapsed time, which gives a negative value. So, slopes going upward to the right have positive slope; that is, lines going upward to the right have positive slope. Lines going down to the right have negative slope. And, a horizontal line has slope zero—that's a place where the car is standing still.

Now we're going to face a challenge of dealing with a part of the curve where the line is not straight; that is, where the graph is not a straight line. This is a case where the car is not proceeding at a steady speed or a steady velocity. The car is changing its velocity. But, how are we going to capture the instantaneous velocity of that car at such a moment?

So let's look at our graph here of the car that was moving at varying speeds and just focus on one little part of this graph, which we'll capture in a different picture, to see how we could specify how fast the car is going at a moment such as this moment. So, here we look at a blown-up picture of just that small part of the graph. We can see that the graph is a curved line, you can see this curved line here, and we're trying now to think about the instantaneous velocity of the car at this particular time. Well remember, our whole analysis of how we computed that instantaneous velocity in Lecture Two. Our strategy was we said where is the car at just a tiny amount of time after the time that we talked about?

In other words, we're looking at the time $p(t + \Delta t)$, and then we subtracted from that the position of the car that the time t we were focusing on. So, this difference of $(p(t + \Delta t) - p(t)$ represents the distance between one point on the graph, and the position of another point on the graph. In other words, at time t, we were at this location on the road; at time $t + \Delta t$, we were at this position on the road. So, the two coordinates of these two points are: For this point, its first coordinate is t, and it's second coordinate is $p(t)$. This point has two coordinates; its first coordinate is $t + \Delta t$, and it's second coordinate is $p(t + \Delta t)$; that is, the position of the car at time $t + \Delta t$. So, the vertical distance here is the difference between its position, the car's position, at the time of t plus Δt minus the car's position of time t, that's the vertical distance. The horizontal is the time Δt, that's the elapsed time. So, the ratio, this vertical

$$\frac{p(t + \Delta t) - p(t)}{\Delta t}$$

change in position, divided by the elapsed time—that ratio was the ratio that we discussed in the definition of the derivative, which was telling us an approximation of the velocity, and that ratio is also the slope of the straight line that we could draw between these two positions of the car.

Now, notice that the curved line—the actual positions of the car— does not correspond to the straight line between those two points. And, this is where the concept of the limiting process that we were introduced to in the definition of the derivative comes into play because, what did we do next? After taking a particular time Δt, we said that's not the best possible approximation we can get. We need to look at a closer time, a smaller Δt, and see what approximation to the velocity is at that smaller Δt. But we proceed with exactly the same analysis. We see where the car was at this time, even closer past the time we're interested in, t; and we look at—we construct yet another small triangle; and we compute the rise over the run of that triangle—and notice that that rise over the run is the slope of a line that is getting even closer to the curved line. There's less curve; there's less distance for that curve to deviate from that straight line when we pick a smaller distance.

In fact, let's do the process of magnifying our picture. When we magnify our picture, notice that the curved line becomes less curved; and by magnifying the picture, let me be very specific on what I mean by that. I mean that we take the horizontal axis and expand it by, for example, multiplying everything by 2; making everything twice as long; and, then, taking the vertical axis and multiplying everything on it by twice as long; so, exactly the same expansion for both the horizontal component and the vertical component. When we make the same expansion, then a straight line will remain at the same slope. So, when we do this kind of magnification, and we go closer and closer and magnify our curve, which is the exact process of taking smaller and smaller Δt's, we find that the curved line begins to look much more like a straight line, and pretty soon becomes indistinguishable from a straight line. The limit—which is the derivative—is the limit as Δt approaches zero, is, therefore, approaching the value of the slope of the tangent line. That is, the line that just grazes this curve; the line that is of the slope that the curve would look like if we continued to magnify that curve, and it began to look more and more like a straight line at a certain slope. Well, that slope is the derivative of the slope at that point, and it's also the velocity of the car at that time.

So, I want to emphasize two insights that are absolutely fundamental to the derivative and its relationship to the graph of a curve. The first thing is that the graph of a curve is—at any point on the graph of a curve—the slope of the tangent line is equal to the derivative at that point. So, we can glance at a curve, and if it's a curve of the position of a car moving on a straight road, and we point to a particular point on that graph, we can estimate the speed of the car by estimating the slope of the tangent line at that point.

The other insight that comes from this is that smoothly curving lines, when looked at very close, look like straight lines. This is an insight. I want to point this out because it's interesting. I sometimes ask my students to take a circle and to draw on their paper a picture of a circle looked at very closely. So, this is an example of a circle. And, often when a student—I ask them this before I've explained this magnification concept, they will say a picture of a circle looked up close looks like this—it looks like a circle, like a curve. But, in fact, the correct answer is to take out your ruler and draw a straight line. That's what a circle looks like very close. But, all of us are familiar with this. This is not an unusual concept. We're all familiar with this idea because we live on the Earth. If we look around us—the Earth is round;—if we look around us on the Earth, it doesn't look as though the Earth is round. Locally, it looks like it's completely flat. That's because a large circle, when magnified, looks, locally, flat.

So, the derivative of a function gives us the slope of the tangent line at each point of the curve. Now, let's go back to our curve and look at various points on the curve and just see what these slopes are at these various points. In this curve, some places we can tell the slope easily. For example, at these peak points, the slope is 0 because the tangent line is horizontal. If we magnified this point very closely, we'd just see a horizontal line. Likewise here; this is a place where the tangent line is horizontal, and we would say the velocity is 0. At a place such as this one, we can draw the tangent line and we can see what the slope is of that tangent line.

So, where the curve is steep, we see that we have a faster velocity and that the velocity is equal to the slope of the tangent line. So, now, what I'm doing here is dragging along the point of the curve, and showing the tangent line at every point as we drag the point along the curve. And you can see that the slopes are varying from being positive and rather steep here; to 0 here; to being negative here; to being 0 here; to being positive, again, over here. So, what we're measuring and looking at the slopes at each point is the actual value of the derivative at each point.

Now, there's more that we can get from this graph of the position function. We can also talk about the acceleration of the car. When you stop at a stop sign and then you go forward, you increase your speed by accelerating. You go from a speed of 0 to a speed of 60 miles an hour by going through intermediate speeds, and you're increasing your speed at some rate. Well, acceleration is talking about how quickly you're changing your velocity. So, acceleration is a derivative of a derivative of the position function. It's the change in the velocity. Well, we can understand that second derivative, that acceleration, by looking at this curve also.

Look here, for example. Here at the point (0,0) we're going rather fast. The slope of the tangent line has a positive and steep slope. Here at 1 it's 0. So, it has declined. So, this means that the second derivative, the change in the derivative, is going from a positive number down to 0. So, the second derivative is negative; it's decreasing. And it continues to decrease as we go over the top of the curve. And, in fact, if the curve is a curve that is concave down, what does that mean? That means that the slope of the derivative, the slope of the tangent line, is decreasing as we move in the positive time direction this way. So, since the slope is decreasing, that corresponds to the second derivative being a negative number. The velocity is decreasing.

Likewise, when we have an upward cupped part of the curve, that corresponds to—and I'm looking at it so you can see the positive axis going in this direction—the slope of the tangent line increases as we move to the right; gets steeper as we move to the right. That corresponds to increasing velocity; positive second derivative. And, increasing velocity is the same as positive acceleration; acceleration meaning an increase in your velocity.

Well, we've actually then seen all sorts of relationships between the graph of a function and the derivative. Here they are; we've encapsulated them in this chart. When the function is increasing, the derivative is positive; when a function is decreasing, the derivative is negative; when the function is flat, the derivative is zero; when the function is concave up, the derivative is increasing; that is, the slope of the tangent line is increasing when it's concave up, which is the same as saying the second derivative is positive. And the physical interpretation of the second derivative, when we're talking about the position curve, is the acceleration. The acceleration is positive. When the function is concave down, that's saying that the derivative is decreasing. But, since the second derivative is the derivative of the derivative—that means that the second derivative is negative. We're really just referring back up here to these previous insights.

So, given a function, given the graph of a function, we can just take a piece of paper and sketch the graph of its derivative by, at each point on the curve, just measuring the slope of the tangent line and then making a point at the value of that slope. So, if we go back to the curve that we've been looking at several times, as we move a dot along this curve, we'll look at the value of the slope of the tangent line; that is, the value of the derivative; and plot it. That is, we just make a dot. So, for example, at this point we plot the point 0 right here. We're plotting the derivative, the slope of the tangent line, at each point. So, it's 0 here; it's 0 here; it's negative here; it's positive here; and, in fact, if we combine these together as we move the dot along the curve, we're drawing—we're capturing the slope of the tangent line and just marking a dot at that value. We'll see that we're drawing a parabola-looking curve, it's a cup-shaped curve and that is the derivative curve. So, the derivative of this curve of motion is a cup-shaped curve.

There's another insight that comes from derivatives that's rather interesting, and this has to do with the global picture of a trip that you might take in a car. This happens if you're driving on the Pennsylvania Turnpike, and you get on the Turnpike at one point and they give you a time-stamped ticket. Now, suppose you drive 100 miles on the Turnpike and you get off the Turnpike and they take your ticket and it's exactly 1 hour later. Okay? So, you've gone 100 miles, it's 1 hour later, and you drive slowly into the Turnpike and they say, "Well, sir, yes you owe the Turnpike thing, and you also owe a fine because you were speeding." And you say, "Well but no, I wasn't speeding, because you saw me just drive slowly into the turnpike thing." But then they say, "Well yes, but the fact that you went 100 miles in 1 hour means that at some point you had to be going at least 100 miles per hour." This is, actually, the insight of the mean value theorem; that if you are traveling along an interval of time, and your average velocity is a certain amount, then at some point during that time you actually were going that average velocity, and we can see this graphically.

So, this is the graph that expresses the mean value theorem. We start at one point in the graph, at a certain time, that is one point on a road at a certain time; we end up at a future time at a different point on the road; and, during the course of that, we have gone different positions on the road. We sometimes went backward; sometimes went forward. And the claim is that at some particular point, and maybe several different times, our instantaneous velocity was exactly equal to the average velocity during the whole period. And the proof is very simple. The proof is that we just draw the line from our beginning position to our final position on the graph; the slope of that line is equal to the average velocity during our trip; and, now, we just take a line that is parallel to that steepness and just let it float down until the first time it hits our curve. And where it hits our curve will be a point where it is the tangent line at the curve. So, at that point, the derivative is equal to the slope of that line which is parallel to the average velocity; and, therefore, its instantaneous velocity is equal to the average velocity.

I wanted to tell you a brief story about a rule about derivatives called L'Hôpital's Rule, and I'm not really going to explain the actual rule of it, but it's a theorem in mathematics that relates a ratio of derivatives to a ratio of the functions. And, with certain conditions, we can see that the limit of the ratio of derivatives is equal to the limit of the ratio of the functions. L'Hôpital was a mathematician who wrote the first calculus textbook. He wrote this book in 1696, and he wrote this book, and appeared this particular theorem—which all students who take calculus these days will see, L'Hôpital's Rule. So, he's become famous throughout time for this rule. Well, it turned out that L'Hôpital did not prove this rule, he bought it. He had a financial relationship with one of the Bernoulli mathematicians that he would take credit for the mathematical results that Bernoulli was able to produce, and one of them was L'Hôpital's Rule, which everybody knows as the rule associated with L'Hôpital. So, I think his investment definitely paid off, whatever he paid for that.

I just wanted to say two words about notation. Newton and Leibniz had different notations for the derivative and, incidentally, for the integral. Newton's notation was that he took a variable, like x, which he thought of, really, as the function of time, and he just put a dot over it. That was the derivative. Now, there was one problem with this notation for the derivative and that is that a fly, if a fly lands at a particular place on the page, can actually take a derivative. That was a bit of a drawback for this notation. But, Leibniz had a notation for derivative that captures the defining quotient property that we've talked about several time; that is, the change in position divided by the change in time, and one writes this whole fraction-looking thing as the derivative. Now, it has its own problems. Namely, it's not really two different quantities; it's the whole fraction that's one idea. So, that's sometimes confusing to people. But this is Leibniz's notation for the derivative, and a very common notation that we'll be using throughout the course is just to put a prime mark after a function to represent the derivative of that function. So, if we have the function $p(t) = t^2$, then we would note $p'(t) = 2t$, and this is probably the most common notation that is used for derivatives today.

In the next lecture, then, we will be talking about the algebraic manifestation of derivatives. I'll look forward to talking to you then.

Lecture Six
Derivatives the Easy Way—Symbol Pushing

Scope:

Much of the practical power of calculus lies in dealing with specific functions that model physical and conceptual situations. We now have a good conceptual sense of what the derivative means both physically and graphically. In this lecture, we'll see how to compute derivatives algebraically. Most functions that are used in physics, economics, geometry, or almost any area of study are expressions that involve basic arithmetic operations—addition, subtraction, multiplication, division—or exponents or trigonometric functions. Here, we see how these functions give rise to neat expressions for their derivatives and how these expressions agree with the geometric properties of the graphs we observed in the previous lecture. If we have an algebraic expression that tells us the position of a moving car, then we can deduce the algebraic expression for the velocity of the car at each moment without having to carry out the infinite process involved in taking derivatives at each point.

Outline

I. In this lecture, we'll look at the derivative as it is manifested in algebra.

   A. Most functions used in almost any area of study are expressions that involve basic arithmetic operations.

   B. Here, we see how these functions give rise to expressions for their derivatives and how these expressions are obtained in a mechanical way.

   C. We will also see how these expressions agree with the geometric properties of the graphs we observed in Lecture Five.

II. Derivatives would be of no practical value if we had to do an infinite process at each point of time. Fortunately, we don't.

   A. The simplest function describing a moving car would occur when the car is moving at a steady velocity.

1. In this case, the position function is $p(t) = ct$, where c is a constant.
2. The velocity, which is the same as the derivative, is just c.
3. Thus, if $p(t) = ct$, then $p'(t) = c$.

B. Let's look at a function we have already seen:

1. $f(x) = x^2$ (which we have seen previously as $p(t) = t^2$).

Function	Derivative
x^2	$f'(x) = 2x$
0.7	1.4
1	2
1.4	2.8
2	4
3	6

2. Notice in the table that for every number x, the derivative value was $2x$. How can we see in general that this is the case?
3. We can check this derivative algebraically by considering the defining quotient for the derivative, namely,

$$\frac{f(x + \Delta x) - f(x)}{\Delta x} = \frac{(x + \Delta x)^2 - x^2}{\Delta x} = \frac{x^2 + 2x\Delta x + (\Delta x)^2 - x^2}{\Delta x}$$

$$= \frac{2x\Delta x + (\Delta x)^2}{\Delta x} = \frac{(2x + \Delta x)\Delta x}{\Delta x} = 2x + \Delta x.$$

4. Now we take the limit as Δx gets smaller and smaller.
5. Whether the function is $p(t) = t^2$ or $f(x) = x^2$, the answer is the same: $p'(t) = 2t$ or $f'(x) = 2x$.

C. Here are some related functions:

1. If $f(x) = 5x^2$, then $f'(x) = 5(2x) = 10x$.
2. More generally, if $f(x) = cx^2$, then $f'(x) = 2cx$.

D. Let's consider other functions of the form x^n.

1. For the function $f(x) = x^3$, $f'(x) = 3x^2$.

2. We can also understand this derivative algebraically by considering the defining quotient for the derivative, namely, $\dfrac{(x + \Delta x)^3 - x^3}{\Delta x}$, and looking at the limit as Δx gets smaller and smaller.

3. We can see the pattern for finding the algebraic formula for the derivative of powers. If $f(x) = x^n$, then $f'(x) = nx^{n-1}$.

E. We see that if $h(x) = f(x) + g(x)$, then $h'(x) = f'(x) + g'(x)$.

1. Likewise, you can take one given function whose derivative you know and multiply it by a constant to get another function: For example, if $h(x) = cf(x)$, then $h'(x) = cf'(x)$.

2. This allows us to take derivatives of such functions as $5x^3 + 2x$.

   a. We see this as a sum of two different functions, $5x^3$ and $2x$, and we see each of those as a product of a constant times a function whose derivative we know. Therefore, if $h(x) = 5x^3 + 2x$, then $h'(x) = 15x^2 + 2$.

   b. And if we add a constant to the same function, for example, $h(x) = 5x^3 + 2x + 3$, then we still have $h'(x) = 15x^2 + 2$. Why? Because the derivative of a constant is 0, so it does not change the derivative.

F. Note that the derivative of a product is not simply the product of derivatives. The product rule and the quotient rule (if you have a function that is the quotient of two functions) are both rather more complicated algebraic equations.

G. At this stage, we can take the derivative of any polynomial, that is, a function of the form $a_n x^n + a_{n-1} x^{n-1} + \ldots + a_1 x + a_0$.

1. For example, let's take the function $f(x) = x^3 - 6x^2 + 9x + 1$. Note that the graph depicting this function starts down at the left, goes up, then down, and then goes up again.

2. The derivative of this function is $f'(x) = 3x^2 - 12x + 9$. Note that in the graph of this derivative, we see that it is the form of a parabola, going down and up again.

III. The derivative reduces the number of "bumps" in the graph of a function.

    A. A typical third-degree polynomial has a graph that goes up-down-up.

    B. Its derivative has a graph that goes down-up.

    C. A typical fourth-degree polynomial has a graph that goes down-up-down-up.

    D. Its derivative has a graph that goes up-down-up.

    E. We can see this geometrically by tracing the moving tangent line and recording the changing slopes.

    F. In looking at these examples, we see that the derivative has one less bump.

IV. Let's consider a function that is defined geometrically on a circle.

    A. If we take a right triangle whose angle is θ, the sine of the angle θ can be thought of as the ratio of the length of the side opposite θ divided by the length of the hypotenuse. The cosine of the angle θ is the ratio of the length of the side adjacent to θ divided by the length of the hypotenuse.

    B. On a circle of radius 1, the hypotenuse is 1; thus, sine of θ is just the vertical coordinate, and cosine of θ is the horizontal coordinate. That is, $(\cos \theta, \sin \theta)$ are the coordinates of the point on the unit circle corresponding to the angle θ.

    C. As θ changes, so does the $\sin \theta$: When θ is small, so is $\sin \theta$; when θ approaches 90 degrees (or $\dfrac{\pi}{2}$ radians), $\sin \theta$ approaches 1, and $\sin\left(\dfrac{\pi}{2}\right) = 1$.

    D. Similarly, when θ is small, $\cos \theta$ is close to 1; when θ approaches 90 degrees (or $\dfrac{\pi}{2}$ radians), $\cos \theta$ approaches 0, and $\cos\left(\dfrac{\pi}{2}\right) = 0$.

E. Notice that the graphs of sine and cosine oscillate because their values repeat each time we move around the unit circle.

F. What is the derivative of sine? Look at the rate at which the line opposite the hypotenuse is changing in relation to a change in the angle.

G. The derivative of sine is cosine, and the derivative of cosine is negative sine.

H. We can graph these functions and see geometrically why their derivatives are related as they are. Neat.

V. An interesting question is this: Does there exist a function that is its own derivative at every point?

A. That is, is there a function $f(x)$, such that $f'(x) = f(x)$ at every x? Equivalently, we are looking for a function y that satisfies $\dfrac{dy}{dx} = y$.

B. It turns out that the *exponential function* $f(x) = e^x$, where $e = 2.718281828...$, satisfies these conditions.

C. This leads us to the following table of derivatives:

Function	Derivative
$f(x)$	$f'(x)$
1	0
x	1
x^2	$2x$
x^3	$3x^2$
x^n	nx^{n-1}
$\sin x$	$\cos x$
$\cos x$	$-\sin x$
e^x	e^x

VI. If we are trying to find the answer to a question that involves derivatives, we will be able to solve it in a practical way.

A. These equations help to mechanize the process of finding the derivative.

B. If a law of nature, for example, involves derivatives, then we will often be able to express that law with a simple, computable formula.

C. We will use this ability in the next lectures that show some applications of the derivative.

Readings:
Any standard calculus textbook, sections on differentiation formulas.

Questions to Consider:
1. Suppose $f'(2) = 3$ and $g'(2) = 4$. If $h(x) = f(x) + g(x)$ for every x, why is $h'(2) = 7$? This shows that the derivative of the sum is the sum of the derivatives. Warning: This pattern does not hold up for products.

2. Why do you believe that many aspects of nature and human creations are so well described by rather simple functions?

Lecture Six—Transcript
Derivatives the Easy Way—Symbol Pushing

Welcome back to *Change and Motion: Calculus Made Clear.* In the previous lecture we saw how the derivative relates to the slope of a tangent line; the derivative is the slope of the tangent line to a graph of a function. But, much of the practical power of calculus lies in dealing with specific functions that model physical and conceptual situations. So, in the last lecture, we got a good graphical sense of what the derivative means, and we talked about the car moving on a road to get a physical sense of what it means, but in this lecture we'll see how to compute derivatives algebraically.

Most of the functions that are used in physics, economics, geometry, or almost any area of study are expressions that involve basic arithmetic operations—addition, subtraction, multiplication, division, or taking exponents, or trigonometric functions, like sines and cosines. So, here, we're going to see how these functions give rise to neat expressions for their derivatives so that we can see how those expressions are obtained in sort of a mechanical way. That's what this lecture is about.

We'll also see how these expressions agree with the geometric properties of the graphs that we observed in the previous lectures. So, the point is that if we have, for example, an algebraic expression that tells us the position of a car moving along a straight road, then we can deduce the velocity of the car by deducing an expression, an algebraic expression, that tells us the velocity of the car each moment without having to go through this laborious process of computing the limit at every single moment; or the process of looking at the graph of the curve and then just estimating the slope of the tangent line. So, this lecture, then, is about the derivative as it's manifested in algebra. And it's one of the areas that most students think of as taking derivatives. What they're really saying is, "If we give you an algebraic expression, can you find the algebraic expression for its derivative?"

So, let's begin, as always, with a simple example, and then move to more complicated ones. Let's begin with this example. Suppose that we have a function, which we could think of as a position function, or we could just think of as just a general abstract function, $p(t) = $ a constant $c \times t$. Notice that if the constant is 1, this will be a 45-

degree line; if it's 2, it'll be a line with slope 2; and so on. We saw that the derivative at each point of such a function is just equal to the constant c because it's equal to the slope of the line at every point; and since it's a line, the slope is the same at every point on that line; and, therefore, the derivative function is a constant c, which has a graph that's just horizontal. So, that's our first and simplest example of a function that has a nice derivative expression.

Now, let's go to a slightly more difficult one, but one that we've dealt with before. Suppose we have a function $f(x) = x^2$. Now, we saw this expression before in dealing with a car moving on a road, and we saw that the way to compute the derivative was to evaluate $\dfrac{f(x + \Delta x) - f(x)}{\Delta x}$. And, then, look at that fraction as Δx became increasingly small and to see if those numbers approached one common limit; and then we called that limit the value of the derivative.

Now, notice, by the way, that in our previous familiarity with this function, we called it $p(t) = t^2$ for position; the position at time t was t^2. And now I've changed it to $f(x) = x^2$. That makes no difference from the abstract concept of a function, it doesn't matter if we're calling the name of the function p or f, or if we call the variable t or x; it doesn't matter. And so, this is just a tiny step in the direction of abstracting the concepts of the car moving on the straight road to now realizing that any two dependant quantities—any function $f(x)$—that may relate any two kinds of dependent quantities, can have the same analysis of taking derivatives.

We saw before—that we made a chart where we had actually gone through the laborious process of computing this characteristic difference and quotient that's associated with the derivative. We saw that for a function $f(x) = x^2$, that the instantaneous—that the derivative at every point were the values in this chart. Namely, at .7 it was 1.4; at 1 it was 2; and in each case, the derivative was exactly $2x$.

If we wanted to graph the relationship between the graph of the function and the graph of the derivative, here is the graph of the function, $f(x) = x^2$, and here is the graph of its derivative, $2x$. So, we see that if we use the perspective of the moving tangent line, we can see that it accords with the graph of the straight line with slope $2x$.

Because, when we look at this left-hand part of the parabola graph, we can see that the slope is a negative number; and, sure enough, the value of the derivative is a negative value. When it gets to $x = 0$, we see that the tangent line is horizontal; that is to say that the derivative has value 0, and here it is, value 0; and, then, when we moved to the positive values of x, we see that the slope of the tangent line is a positive number and increasingly steep, and that accords with this derivative function having slope 2 and continuing to rise as we get larger values of x.

So, once again, we see an algebraic expression for the derivative. But, why is that algebraic expression true? Let's go ahead and do the actual mathematical derivation of the derivative. Remember that the derivative, $f'(x)$, is equal to—and I'm now going to actually use the notation associated with taking the limit. That is it's expressed by saying LIM for limit, as Δx approaches 0, with that arrow; Δx with an arrow towards 0 means that we look at this expression that follows that limit symbol, namely, $\dfrac{f(x + \Delta x) - f(x)}{\Delta x}$, and evaluate that difference quotient for smaller and smaller values of Δx, as we choose Δx to be increasing small and getting closer and closer to 0. We evaluate that quotient and see that those numbers get closer and closer to one value.

Well, let's see what value they get closer to. So, the limit as Δx approaches 0 of this expression, well, what is $f(x + \Delta x)$? We're looking at the function $f(x) = x^2$. So, $f(x + \Delta x) = x + \Delta x^2$. And, the function of x is x^2. So, we have—this is the numerator, and then Δx remains as the denominator of that fraction; and, then, just doing the expansion of this expression. Just in other words, multiplying out, using the distributive law, we can see that $(x + \Delta x) \times$ itself is equal to x^2

$$p'(x) = \frac{p(x + \Delta x) - p(x)}{\Delta x}$$

$$= \frac{(x + \Delta x)^2 - x^2}{\Delta x}$$

$$= \frac{x^2 + 2x\Delta x + (\Delta x)^2 - x^2}{\Delta x}$$

$$= \frac{2x\Delta x + (\Delta x)^2}{\Delta x}$$

$$= 2x + \Delta x$$

$$= 2x, \text{ as } \Delta x \text{ becomes small}$$

+ $2x\Delta x + (\Delta x)^2 - x^2$, all over Δx. The x^2's conveniently cancel; leaving $2x\Delta x + \Delta x^2$. The Δx's can be factored out, so that we have in this expression $2x\Delta x + \Delta x$, divided by Δx. The Δx's cancel, leaving the limit as Δx approaches 0, which, of course, has preceded in each of these lines, of $2x + \Delta x$. All that means is that if we take any value of x, and we plug in any value Δx that's not 0, but maybe close to 0—it would be true even if it weren't close to 0, but we're interested in those values that are close to 0—if we plug it in, all of this algebra is just telling us that the value of this quotient is exactly just $2x + \Delta x$. That's what it says.

Now, what happens as Δx becomes increasing small, increasingly close to 0? Well, this expression, then gets increasingly close to $2x$, and then in the limit it's equal to $2x$. That is the derivation of the formula for the derivative of the function of $f(x) = x^2$. Now, this is a good example of the kind of derivation involved in actually finding algebraic expressions that equal the derivative of a given algebraically expressed function. So, I hope that you enjoyed that.

Now, let's look at some related functions and just see how we get their algebraic derivatives. We just saw this one. Suppose that we took our expression, instead of x^2 we multiplied it by some number, like 5. What is the result? Well, it's easy to see by just looking at the derivation and just thinking about what it means to multiply by the constant 5; that means that the values in the vertical direction are all expanded by a factor of 5, which means that every single quotient will be five times as tall on that vertical part of the triangle as it was in the previous derivation. And consequently, the value of the derivative is going to be five times as great as the value of the derivative before. And in general, there is the general statement that if you have $f(x) =$ a constant $\times x^2$, the value of the derivative—and you can think of it as the slope of the tangent line—is always going to be just $2 \times$ that constant $\times x$. In other words, the constant just flows through the taking of the derivative.

Let's do one more example of taking a derivative to get a spirit of how these go for functions of the form x to a power. So, this one is the function $f(x) = x^3$; and we're going to just follow the exact form that we did before; namely, we realized that the definition of the derivative of that function, $f'(x)$, as always, is the limit as Δx approaches 0 of the function evaluated at $x + \Delta x$ – the function evaluated at $\dfrac{x}{\Delta x}$. When you think of derivative, you should always, immediately, come to this expression; this characteristic fraction.

Well, what is the function of $x + \Delta x$? Well, the function we're talking about just takes any value and cubes it to give us the value there, so we just cube $x + \Delta x$. This is cubed $-\dfrac{x^3}{\Delta x}$, multiplying it out. Once again, the x^3's cancel; we're left with some expressions, each of which has a Δx in it. After we factor out the Δx's and cancel the denominator, we're left with only one expression that fails to have a Δx. The two other expressions, $3x\Delta x$ and $-\Delta x^2$, are expressions each of which contains a Δx in them. Now, notice, as Δx goes to 0, that's what the limiting process is. It says what happens when you take Δx to be .0000001, these expressions here are going to become negligible while this expression stays the same because it

$$, \quad = \frac{p\ x + \Delta x\ -}{\Delta x}$$

$$= \frac{(x + \Delta\ -}{\Delta x}$$

$$= \frac{3^{\ 3} + 3^{\ 2} \quad x(\Delta x)^2 - (\Delta x)^3 - x^3}{x}$$

$$= 3x^2 + 3 \quad \Delta \quad (\Delta x)^3$$

$$= 3x\ ,\ \text{as}\ \Delta x\ \text{becomes small}$$

doesn't have a Δx in it. So, if x is equal to 5, for example, this would stay 3×5^2—it would be 75—plus some value that if I chose little tiny Δx's, those would become .0000001; they become negligible, and in the limit, disappear. Consequently, the derivative of $x^3 = 3x^2$.

Well, maybe we're beginning to see a pattern here; and the pattern is so far, the derivative of x—remember, that has slope 1, so its derivative is 1; the derivative of x^2 is $2x$; the derivative of x^3 is $3x^2$; and the pattern continues. In general, for any function $f(x) = x^n$, its derivative is equal just to $n \times x^{n-1}$ power. And this algebraic way of taking derivatives becomes very habitual for students. In fact, many of them believe the derivative is taking down the exponent, multiplying and subtracting 1 from the exponent, without—failing to remember that the derivative is telling us something of meaning;

mainly, the slope of the tangent line, or the velocity of a car moving on a road, if the function is telling us the position of the car on this straight road on every moment. So, we have our chart of derivatives for exponential—for functions that are just x to a power.

Much of the way that we develop our strength of taking derivatives of more complicated functions is to see how, if you can get a function as being constructed from other functions by various operations, such as adding two other functions together to get a new function or multiplying two functions together to get a new function, that we want to see how the derivatives of the newly created function relate to the derivatives of the individual component part functions. So, we'll start that process by talking about the sum of two functions. In other words, suppose that we have a function $h(x)$ that equals the sum of two other functions, $f(x)$ and $g(x)$. In other words, to get the value for a function h, you first compute the value of f and you get a number; you compute the value of g at that same point, you get a number; you add them together to get the value of h. Then, the derivative, as you would expect, is just equal to the derivative of f plus the derivative of g, at that point; at every point.

Similarly, let's take another way that you could construct a function. You could take one given function, whose derivative maybe you know, and multiply it by a constant to get another function. We already saw that in the case of a constant times x^2, and it more generally applies to any function; that is to say that the derivative of a constant times a function, that is, if we create a new function by taking an old function, $f(x)$ and multiplying everything by a constant, then the derivative of that new function is just the constant times the derivative of the old function, for the same reason that we talked about before, just the vertical components being expanded by the factor c.

This allows us to take derivatives of functions such as this one: $5x^3 + 2x$; because, you see, we see this as a sum of two different functions, $5x^3 + 2x$, and each of those is the product of a constant times a function whose derivative we know; like, the derivative of x is equal to just 1, so 1×2 is 2; and the derivative of x^3 is $3x^2$; and, consequently, $5x^3$ will have a derivative that is five times as big as the derivative of x^3; so, it's just $3 \times 5 = 15x^2 + 2$. So, this is the way of understanding the derivative of a function that is created by the sum or a constant times a given function whose derivative we already know.

Notice this slight variation on this. Let's look at a function that's the same function except that we add a constant. What's the derivative of a constant? Well, the derivative of a constant, a constant has a graph that is just horizontal, that derivative is 0 at every single point; and, consequently, it does not change the derivative. The derivative is exactly the same if we add a constant. By the way, the look of a func

$$h(x) = f(x)g(x) \text{ then } h'(x) = f'(x)g(x) + f(x)g'(x)$$

tion, the

graph of this function compared to the graph of this function, all we've

done is shifted the graph up

$$h(x) = \frac{f(x)}{g(x)} \text{ then } h'(x) = \frac{g(x)f'(x) - g'(x)f(x)}{[g(x)2]}$$

by three units. That's what it means to add a constant. And, since we've just shifted the graph up by three units, the tangency at the same point is exactly the same; and, since the slope of the tangent is what the derivative is measuring, we get the same answer when we add a constant.

When it comes to products of functions, we have a surprise. You might think that the derivative of the product of two functions is the product of the derivatives, but this is false; this is simply false. One good way to realize it must be false is the derivative of x^2—x^2 is $x \times x$—but it's not equal to the derivative of $x \times$ the derivative of x. The derivative of x is 1; so, it's not 1×1; in fact, it's $2x$. So, that is not—that shows the simple product rule, the naive one that you might think of, is not correct. The actual product rule is the following—it's a little more complicated—that if you construct a function, $h(x)$, by taking the product of two previously known functions, $f(x)$ and $g(x)$, then the derivative of the product is the first times the derivative of the second plus the second times the derivative of the first.

Likewise, there's a quotient rule. If you have a function that's the quotient of two functions, then its derivative is a rather complicated expression. It's the denominator function times the derivative of the numerator function minus the derivative of the denominator function—the one downstairs—times the numerator function divided by the square of the denominator function.

Now, each of those can be deduced by just writing down the definition of derivative and then doing some algebra.

At this position, now, at this stage, we can take the derivative of any polynomial. A polynomial is a function of the form something times x to a power plus some other constant times x to another power plus something times x to another power. Let's look at a specific example. Suppose we look at the example $f(x) = x^3 - 6x^2 + 9x + 1$. The graph of this function, as you see, it starts down to the left, it goes up, it comes back down, and back up again. Notice, actually, this is exactly the same function that we used in the previous lecture. Suppose that we take its derivative; that is, we actually compute the derivative using now the formula rules that we've deduced. The derivative of $f(x) = 3x^2 - 12x + 9$; and, if we graph that derivative on the same graph as we see the graph of this function, we can see that it corresponds the way we saw visually that it must correspond in the previous lecture. Namely, at this peak point of the original function, that is the point on top where the derivative is—where the tangent line is horizontal, that is to say the derivative is 0. Well, we notice that the derivative function is equal to 0 at that point. Likewise, at this minimum value, the point where it's at the bottom of this trough; that is, the graph of the function has this trough. Once again, the tangency is horizontal at that point; that is, the derivative has value 0; and, sure enough, the value of the derivative function at that point is 0 as expected, as it must be. And also notice that the graph of the derivative function has just—it's called a parabola; it just goes down and up again. It just has one change in direction; whereas, the function that we started with, was a cubic function. It went up, down, and up again. It's characteristic of polynomials that higher-degree polynomials have more oscillations, and that when we take the derivative, the number of their oscillations declines by one. I always find this an interesting thing, that you can actually graphically see that the number of oscillations will decrease by one, and it is expressed algebraically as well. It's manifested by the algebra and by the geometry together.

Here's an example of a fourth-degree equation, and we take it's derivative and draw the graph of the derivative on the same graph, and we see once again that the number of this fourth-degree equation goes, in this case, down, up, down, and up again; and, then, the derivative function, which is a third-degree polynomial, starts low because you see here, the derivative has—the tangent lines are negative and steeply negative. So, the value is down here, and then it goes up, down, and up again. So, once again, we have a manifestation that the derivative, the formula for the derivative, is giving us something that has geometric significance as well.

Now we're going to turn our attention to two additional classes of functions that are of interest. The first one is the trigonometric functions. Now, the trigonometric functions, particularly the sine and the cosine, are two of the most fundamental of the trigonometric functions; and let me, first of all, remind you how the trigonometric functions are defined. The sine of an angle Θ can be thought of as the quotient of the—if we take a right triangle whose angle is Θ, it's the ratio of the opposite side of Θ divided by the length of the hypotenuse. That ratio is the sine of Θ, and the cosine of Θ is the adjacent side divided by the hypotenuse.

Another way to talk about such functions is the following: Let's just take a unit circle, so that the hypotenuse has value 1; then all of the points on the unit circle have two coordinates. The first coordinate is the cosine of the angle that's measured by starting at the horizontal axis and measuring counter-clockwise around to get to the angle that we're talking about. The x-axis is equal to the cosine, actually, of Θ, because it's the adjacent side over 1. So, this distance here, which is the first coordinate, is the cosine of Θ. This second coordinate is the sine of Θ. So, the cosine Θ, sine Θ are the two coordinates of a point on this unit circle. By the way, for convenience, we measure this angle Θ by the arc length around the circle, of the curve. So in other words, as we go along the curve, remember a unit circle has circumference $2\pi \times 1$, so just 2π; and so this measure Θ goes from 0, for example, up to this distance, it's $\pi / 2$; that's a quarter of the way around; halfway around is π; then $3\pi / 2$; and, then, 2π, after you've gone once around—and notice that if we go twice around, we get the same value as not going anywhere. Every time we go 2π or more around, we get the same value.

Now, the sine and the cosine, you see, oscillate because every time we go 2π around, they oscillate between what they started with, and then they come back to that same value. So, if we graph the sine and the cosine, we'll see that the sine varies between 0, at $\Theta = 0$; it gets up to 1 at $\Theta = \pi / 2$; it goes back to 0 at π; it goes to -1 here; and then it gets back to 0 here; and then it oscillates. So, the graph of the sine function goes up and down and up and down, oscillating forever.

It turns out that the derivative of the sine function is the cosine function. It oscillates also, but it starts at 1 and it goes down to—the value of the cosine at 0 is 1 because, you see, the horizontal coordinate is 1, and 0, and then it oscillates going back and forth, and you can see just graphically that as we slide the tangent line along the sine curve, the slope of that tangent line is 1, and that's the value of the cosine at 0. Likewise, when the sine reaches its maximum at 1, here at $\pi / 2$, the maximum value of 1 at $\Theta = \pi / 2$; the cosine is down at 0. And, so, those two things oscillate together in making the derivative, the cosine is a derivative of the sine. So, here we can summarize this that the derivative of the sine is the cosine, and the derivative of the cosine is negative of the sine function.

I wanted to talk about one other class of functions, and these are interesting functions that are exponential functions. Could we have a function whose derivative is exactly equal to itself? Well, that's an interesting question. Suppose we started at the point 1 and we said could we draw a function whose derivative is exactly equal to the value of the function? Well, if we started our function, just arbitrarily, at value 0, having slope equal to 1; in other words, it starts up at a 45-degree angle, then we could start drawing the function by just saying well, it goes up by about 45 degrees, but then if we want its derivative to be equal to itself, then since the value is increasing of the function, you see, because it's going up, then the derivative has to be steeper. So, it has to become steeper, and it has to become ever steeper as it goes up, and that creates an exponential function. And, in fact, we can find such an exponential function just by drawing it. It has the name e^x; and e^x is a function with the interesting property that its derivative is equal to itself. And, in fact, it's equal to the function where there's an actual number associated with it if you take the number 2.718281828459.... you will find that when you raise it to a power and you actually compute the derivative, you will find that the derivative at every value is equal, exactly equal, to itself. Of course, the actual number goes on forever, it's not a number that stops at any given point.

Exponential functions in general have the property that the derivative of taking any constant number raised to a power, like 2^x power, by looking at this algebra, the definition of the derivative, we see that since we can factor out the a^x itself, we see that, in fact, the derivative of any exponential function, a constant to the x power, is that same value times a constant. This leads us to a table of derivatives. So, so far we have this following table of derivatives; we know how to take the derivatives of x^n power; the derivative of the sine is the cosine; the derivative of the cosine is minus the sine; and the derivative of e^x is e^x.

So, the purpose of this lecture was then to give us a mechanical way for taking an algebraic expression that may express some relationship in nature or in science or in economics, and if we can express it in some relation using the functions that we have, then we can automatically understand what the derivative of those functions are in this mechanical way.

In the next lecture, we'll see the derivative in many different application areas. I'll see you then.

Lecture Seven
Abstracting the Derivative—Circles and Belts

Scope:

The true power of the derivative lies in its ability not only to help us understand change with respect to time but also to deal with totally different types of dependent quantities. For example, the area of a circle is dependent on its radius; thus, we can describe the rate at which the area changes when the radius changes. In the previous lecture, we saw how the derivative behaves algebraically, so we can easily compute algebraically the derivative of the function that tells us the area of a circle of a given radius. Here, we show how to interpret the answer geometrically. That is, we see what the derivative means visually when we apply the concept of derivative to formulas for area and volume. Our geometric insight leads to a surprising realization about the roominess of belts around the Earth.

Outline

I.   In this lecture, we continue our study of the derivative.

    A. In Lecture Five, we saw how the derivative is expressed graphically.

    B. In the last lecture, we saw how to express a derivative algebraically.

    C. In this lecture, we will see how to interpret a derivative geometrically.

II. The concept of the derivative can be generalized to apply to any two dependent quantities.

    A. The derivative can be applied to all cases of two interrelated quantities.

        1. The derivative describes how a change in one quantity entails a change in the other.

        2 For example, the area of a circle is dependent on its radius. We can discuss quantitatively the rate at which an increase in the radius of a circle would change the area of the circle.

B. The derivative measures the rate at which a change in a variable causes change in a dependent quantity.

III. We will now see that the algebraic formulas for derivatives developed in the previous lecture make perfect intuitive and geometrical sense in situations other than motion, as well.

A. We start with the square. The area of a square is $A(x) = x^2$, where x is its side length.

1. Algebraically, we know that the derivative is $A'(x) = 2x$. That is, for every unit increase in the side length x, the area increases by roughly $2x$. But where does this come from?

2. Let's see visually what the derivative means. That is, we want to compare the area of one square with the area of another square that is slightly bigger. Traditionally, we use the term Δx to measure the change in the length of the side.

3. At a given side length, we can compute how much the area would increase if we increased the length of the side.

4. The influence of a change in the length of the side varies depending on how big the square is to start with.

5. Note the fraction we get when we look at the ratio of the change in area over the change in the length of the side:

$$\frac{(x + \Delta x)^2 - x^2}{\Delta x}$$

6. We find the exact rate of change by taking the limit as Δx goes to zero and arriving at the derivative.

   a. For the larger square, we take the augmented rectangular area of one side, which is $x\Delta x$, and add it to the augmented area of the other side, $x\Delta x$, and finally, we add the area of the smaller square where the added lengths meet $(\Delta x)^2$.

   b. We then divide by Δx, so we have:

$$\frac{(x\Delta x + x\Delta x + (\Delta x)^2)}{\Delta x} = 2x + \Delta x.$$

    c. As Δx gets smaller and approaches 0, it becomes negligible, and essentially, we arrive at $2x$, exactly as predicted by the algebra.

B. Let's now consider the volume of a cube. The volume is $V(x) = x^3$, where x is its side length.

    1. Algebraically, we know that the derivative is $V'(x) = 3x^2$. That is, for every unit increase in the side length x, the volume increases by roughly $3x^2$. But where does this come from?

    2. As with the square, at a given side length, we can compute how much the volume would increase if we increased the length of the side.

    3. Let's look at how much volume we add if we increase the length by Δx on all three sides of the cube.

    4. The influence of a change in the length of the side varies depending on how big the cube is to start with. In any case, it adds a layer or "slab" of extra volume on three faces of the cube.

    5. The added volume for one side is $x^2\,\Delta x$, where Δx indicates the added thickness. We do the same for all three sides and arrive at $3x^2\,\Delta x$. To find the rate of growth, we divide by Δx and arrive at $3x^2$. We have four other tiny additional pieces of volume right at three edges of the cube and at one corner, but as Δx grows smaller and approaches 0, these tiny parts of the volume become increasingly negligible and tend to disappear, and we see the rate of growth is $3x^2$, exactly as predicted by the algebra.

C. Let's consider how the derivative would be defined in the case of the area of a circle viewed as dependent on its radius, $A(r) = \pi r^2$.

    1. Again, from the previous lecture, we know $A'(r) = 2\pi r$; that is, for every unit increase in the radius, the area increases by roughly 2π, or a little over 6 units. Why does this make sense?

    2. At a given radius, we could compute how much the area would increase if the radius were increased a certain amount Δr.

3. The ratio of that change in area divided by the change in radius Δr would yield a rate of change of the area with respect to a change in the radius.
4. The additional area (called an *annulus*) is in the shape of a thin washer or ring or belt whose width is Δr and whose length is the circumference of the circle, $2\pi r$.
5. Thus, the additional area is approximately equal to the thickness of the ring times the length around the circle; that is, the additional area is about $2\pi r \Delta r$.
6. When we divide that change in the area by the thickness of the ring, Δr, we get a number that is approximately equal to the circumference of the circle, $2\pi r$.
7. This ratio of change of area divided by change in radius again approximates the influence of radius on area.
8. Thus, we obtain an intuitive understanding for $A'(r) = 2\pi r$.

IV. Suppose a belt is put around the equator of the Earth (about 25,000 miles). Suppose we increase the length of the belt by 7 feet so that it hovers above the Earth. How far away from the Earth will it be?

A. The circumference of a circle of radius r is $2\pi r$.

B. From the previous lecture, we know that the derivative is 2π. This means that every unit increase in r results in roughly 2π (or just over 6) units increase in the circumference.

C. Most people guess wrong about the belted Earth, assuming that the longer belt will hover just slightly off the ground.
1. In fact, the derivative tells us that adding 7 feet to the circumference of the belt will result in the belt being more than a foot away from the Earth!
2. In other words, we have added more than a foot to the radius of the belt. We can understand this result better if we look at the derivative, 2π.
3. A 1-foot change in the radius of the belt brings about roughly an addition of 6.28 feet to the length of the circumference of the belt. Therefore, adding 7 feet to the length of the belt would indeed result in the belt being more than a foot away from the Earth.

V. Now let's consider a stalactite growing in a cave. How quick does a 20,000-year-old stalactite acquire volume compared to its 10,000-year-old cousin?

A. Depending on the environment, stalactites usually grow in cone-like shapes at the rate of 1–4 inches every 1,000 years. We'll assume our stalactites grow at the rate of 3 inches every 1,000 years.

B. We'll also assume that the ratio of the stalactite's base radius to its height is 1:6.

C. If a stalactite is x thousand years old, its height is $3x$ inches and its radius is $\dfrac{x}{2}$ inches. Therefore, at 10,000 years, the stalactite is 30 inches high and has a radius of 5 inches; at 20,000 years, the stalactite is 60 inches high and has a radius of 10 inches.

D. The volume or weight of the stalactite when it is x thousand years old is $W(x) = \dfrac{1}{3}$ base times height, or

$$W(x) = \frac{1}{3}\pi\left(\frac{x}{2}\right)^2 3x = \frac{\pi x^3}{4}.$$

E. What is the growth rate of the weight of the stalactite? It is precisely the derivative of the weight.

F. Using our table of derivatives, we see that $W'(x) = \dfrac{3\pi x^2}{4}$. Therefore, the derivative of the weight at 10,000 years is 75π, while at the 20,000-year mark, it is 300π.

G. Thus, even though the 20,000-year-old and the 10,000-year-old stalactites increase in *height* at the same rate, the older one increases in weight at 4 times the rate of the younger one!

VI. Let's consider how we would define the derivative for dependencies involving supply and demand.

A. We see that the supply curve shows how many items producers would produce if the price they could sell the item at were p. The demand curve shows how many items consumers would buy if the price were p.

B. In this graphical representation of the curves, notice that if the price increases, the supply increases.

C. Also notice that if the price increases, the demand decreases.

D. The place where the demand curve and the supply curve cross is the point where we expect the price to be.

E. What does the derivative measure, when applied to the supply and demand curves?

F. If the curves are steep near the crossing point, slight increases in price would cause major increases in supply and major decreases in demand. These are elastic goods.

G. Goods that are insensitive to price, for example, gasoline, are called inelastic.

VII. We have now seen the importance of the derivative.

    A. The derivative analyzes change.

    B. The derivative gives a quantitative view of how one quantity will change when another quantity on which it is dependent changes.

    C. Analyzing dependencies and change is one of the most fundamental things we do to try to understand our world.

    D. The derivative in general arose from first analyzing the simple case of a moving car. That example illustrated one example of dependency—position was dependent on the time.

    E. We then saw that the same analysis would serve in many cases where one quantity is dependent on another.

Readings:

Any standard calculus textbook, sections on applications and interpretation of derivatives. (These explanations may be difficult to find in standard texts.)

Questions to Consider:

1. Consider an animal whose weight varies according to height according to the function $w(h) = h^3$. Suppose the strength of the

animal's legs at a given height is given by $s(h) = h^2$. Could similar animals exist that are 10 times as tall? Does this help explain the limit to the size of land animals?

2. Suppose the population increases by 1% each year. Does the derivative of the population function (that is, the function whose value is the population at each time) increase, decrease, or stay the same over time?

Lecture Seven—Transcript
Abstracting the Derivative—Circles and Belts

Welcome back. As you recall, we're in the midst of a sequence of three lectures about the derivative and seeing the manifestation of the derivative in many different settings. In the last two lectures, we saw the derivative first geometrically, associated—that is, graphically associated with a graph and the fact that it records the slope of the tangent line at each point of a graph. And, then, in the last lecture we looked algebraically how if we're given an expression, an algebraic expression of a function, we can compute an algebraic expression for its derivative.

Well, the true power of the derivative lies not only in its ability to help us understand change with respect to time, like the car moving down the straight road, but also to deal with dependencies of all different types—whenever we have one quantity dependent on another, that's a place where the derivative can come into play. For example, the area of a circle is dependent on its radius; and we can write a description of an area in terms of its radius, a familiar formula for the area of a circle. And then we can discuss the rate at which the area of the circle changes when we change the radius. That's an opportunity for the derivative to give us some insight.

In the previous lecture, remember, we saw how we can compute things algebraically, and so we can easily compute algebraically a formula for how the radius of the circle is going to influence the area of the circle; that is, a change in the radius influences the change in the area. So, what we're going to do today is to look at some applications of the derivative and think about the geometric manifestations of them, first, and then in some more abstract settings at the end of the lecture.

So, the first one we're going to do is start with the square. Now, the first thing to think about with a square is we're going to remind ourselves of the derivative formulas that we deduced in the last lecture, and, in particular, this one here, that the derivative of $x^2 = 2x$; meaning that if we draw the graph of x^2 for example and we focus our attention at any point x and we go up to the graph, we'll find that the slope of the tangent line at that point is exactly $2x$.

Well, let's think about the area of an actual square. So, here we have an actual square; and the area of the square is the length of one side times the length of the other—they're the same, so it's just $x \times x$. So, when we think of x^2 as meaning something, geometrically, it means that if we have a length x and we construct an actual physical square whose sides are each x, then the total area of that square will be x^2.

Well, let's see what the derivative means in terms of this geometry. I find this sort of interesting, by the way, that if you take this square and now let's go ahead and do the derivative procedure. What is the derivative procedure? The derivative procedure is to say that we take—we want to compare the area of square with the area of a square that's slightly bigger; and, traditionally, we use the term Δx. In other words, we're saying what is the area of a slightly bigger square? Now, this is a square each of whose sides is $x + \Delta x$. So, here we have this bigger square. Now, let's look at the definition of the derivative. The definition of the derivative says that we look at the value of the function at the bigger moment, $x + \Delta x$, minus the value of the function at point x, and divide by Δx. And, then, take the limit as that fraction goes—as Δx goes to zero—we look at that ratio of the difference between the bigger, from the original one, and then divide by Δx and see how that ratio behaves.

Let's interpret that ratio in a completely physical way. Here we have this larger square. So, on our graphic we see that the larger square has sides $x + \Delta x$, and then we're going to subtract the area of this bigger square, from that we're going to subtract the area of the original square we started with. And, we can see geometrically that the area, the augmented area, the additional area, is shaded here; it is the area that sits on these two sides of the square.

Well, now let's see if we can understand the idea of taking this additional area and dividing by Δx. Well, let's see, the area of this rectangle that goes up to the top here and down, it has width Δx, and we're going just up to the top of the original square. So, this height is x; the width is Δx; so, the area of this square right here, this rectangle right here, the area of this rectangle right here, is $x \times \Delta x$. Likewise, the area of this rectangle up here is $x \times \Delta x$. Those two rectangles encompass most of the additional area. Then we have this little square up here in the corner that's also added. It's $\Delta x \times \Delta x$.

Notice that when we divide by Δx, we divide this additional area by Δx, well, if we take this rectangle and divide by Δx; we take the area of that little thin rectangle and divide by Δx, what do we get? We get x. When we take this rectangle, which is—the area of this rectangle, it's $\Delta x \times x$; when we divide by Δx we simply get the length, x. So, what we're getting when we take this additional area and divide by Δx, we're going to get three contributions to that difference quotient. One of them is going to be x, contributed by this rectangle; another one is going to be x, contributed by this rectangle; and then we have this $\Delta x \times \Delta x \div \Delta x$, which is just Δx, which is getting tinier and tinier and becomes negligible as Δx becomes zero. But whereas, these two—the area of this one divided by Δx—for any Δx—will continue to give us this value of x.

The point is that the derivative of the area function can be viewed geometrically. It's saying that if you take a square and you imagine that square growing in size. So, we're imagining this square getting bigger and bigger. You see how we see this square growing bigger and bigger and bigger, that the rate at which the area is increasing is going to be equal to, geometrically, this length plus this length. That's the rate, meaning that when we take those lengths times Δx, that's going to be a good approximation to the additional area that we're creating. So, it's the rate of growth of this square.

So, I find this interesting because it shows that the derivative of the area formula, by viewing this dynamically, you see, it's this dynamic vision of the way we think of this extremely static-looking square, but if we think of it as a growing object, starting at this one corner and growing up bigger and bigger squares, the derivative is telling us the rate at which the area is growing. And, we see it geometrically by the length of those two lines.

So, this is the first example of a geometric interpretation of the meaning of the derivative. Let's do another one just to get in the swing of it. Going back to our formulas for our table of derivatives, we saw that the derivative of x^3 is $3x^2$. We saw algebraically why that's true. We actually went through the derivation and saw that that is the derivative. Let's interpret it once again geometrically. Well, the formula x^3 is referring to the volume of a cube whose sides are x. So, here is an example of a cube. If its side is x, all three of the lengths are x in every direction, then the volume of this object is $x \times x \times x$. That's the volume of a cube. Now, can we envision in our minds what the derivative of x^3 means with respect to this geometrical manifestation of the function x^3?

Well, what does it mean? It means that if we think of this cube as growing; that is, this is a small cube, suppose we had a bigger cube, a bigger box. We could imagine this cube growing from one corner to become a bigger cube. Now, as it's growing, the sides, each of the sides is growing. It becomes longer. And, we're going to ask ourselves, what's the rate at which the volume is increasing? Well, let's look at it geometrically and see the rate at which the volume is increasing.

Well, if you take a cube, and you think about anchoring the cube at one corner, and having the lengths of the three sides increasing, then we see that when we add—we start at a given length x and add Δx to that length; we add Δx to the height; we add Δx to this length coming out the back, we can see that the additional volume, the change in the volume, is a layer of material on three sides of the cube. It adds three layers on the side of the cube: a slab on this face, a slab on this face, and a slab on the back face. Now, we're adding that amount of material, but then to see the rate at which the volume is increasing, we divide by Δx the amount of increase. Well, what happens when we divide by Δx? Let's look at the slab on this face of the cube. That face has area $x \times x$— x^2 —and then it adds the thickness Δx; when we divide by Δx, we simply get the area of the face of that cube, x^2. Likewise, on the top face, doing the same procedure, the volume of this additional material just on the top is $x \times x$ — x^2 — $\times$ the thickness, Δx. When we divide by Δx, we just get x^2 again; and the same for the back.

Now, notice there are three other—or four other pieces of volume that we have not accounted for in this geometric visualization of the derivative process. The three additional pieces of volume occur right along this edge, right along this edge, right along this edge, and then right in the far corner, the diagonally opposite corner. Notice the part on this edge is a little—it's actually a square times x; and on this side, a square times height x. It's a little tiny parallel pipe; a very thin box. We have three of those, and then we have a little tiny cube here in the corner, which is $\Delta x \times \Delta x \times \Delta x$. Notice that as Δx goes to zero, this proportion of the volume becomes increasing negligible; and, when we divide by Δx, those things disappear. The main contribution to the volume is occurring on these three faces. The fact that there are three faces, each of which have sides x^2, is telling us that the rate at which the cube is growing in volume is $3x^2$. So, notice that we have once again seen a correspondence between the derivative formula that we previously had deduced from the last lecture; that the derivative of $x^3 = 3x^2$, and now we've seen physically what that means. What does it mean, $3x^2$? It's the three faces that are growing when the cube is growing.

Okay, so this is the case of a cube. Let's go ahead, since we're doing these geometric objects, let's go ahead and do the circle. The area of a circle, as you know, is πr^2. Now, what that means is that if you take a circle, such as this one, and you measure the radius of the circle, and then you just take that radial measurement, whatever it is, you square it, and multiply it by the number π, which is 3.1415926… and so on, which is the ratio between the diameter of a circle and its circumference. πr^2 will automatically give you the exact area inside this circle. That's what it means to have a dependency of the area on the radius, πr^2. So, you don't need to take each individual circle and measure how many square inches it has; you can just find out how many inches the radius is, and then do this computation, π times that radius times itself, πr^2.

Let's see geometrically what—well, let's see, first of all, algebraically; let's remember algebraically what the derivative of that area formula is. The area formula is πr^2, and we saw in the last lecture that the derivative is obtained by bringing down this exponent 2, the constant π —π is just a constant number, and so it's preserved in the derivative. The derivative of the formula that gives us the area of the circle as a function of the radius, that derivative is $2\pi \times r$. So, that we saw algebraically; let's look at it geometrically and see what it's telling us.

Well, if we have a circle of radius r, and then we increase that circle—we take the circle of radius r, so it has a certain area. Then, we increase the radius by Δr, so we increase the radius by Δr; and we're trying to see the rate at which the area of the circle is increasing as a function of the rate at which the radius is increasing. In other words, for a unit increase in the radius, how much increase in area do we expect to have happen? Well, let's look at it geometrically. Geometrically, what we're saying is that if we look at a circle that's a slightly bigger circle than our circle of radius r. It's slightly bigger by having a radius of $r + \Delta r$. Then that circle just circles around, it's slightly bigger than the original one, and notice that the difference in area, the additional area, is what's called an annulus; it's like a belt, whose width is Δr, and then whose length goes around the circumference of the circle. So, an approximation of the area of that additional area, an approximation would happen if we cut it and thought about straightening it out. Now, of course, you couldn't actually straighten it out because the inner length would be not quite the same as the outer length. The inner length, here, you see, is smaller, shorter, than the outer length. So it wouldn't quite be straighten-outable, but if you did that you would get an approximation to that area in that annular region.

Now, what's the process of the derivative? It's saying the rate at which that area is growing. So, we take that additional area and divide by Δr—that's the change in the radius—to get the rate; that is, the amount of area per added radius. Well, if we think about the approximation of this area as the circumference of the circle, $2\pi r$, that's the inner distance around this circle, times Δr, that's the—so, this is an approximation to the additional area; when we divide that by Δr to get the rate at which the additional area is getting constructed, we simply get $2\pi r$. So, once again, we see that there's a relationship between the formula for that boundary of the circle and the derivative of the area function. So, there's a correspondence between the geometric interpretation of the derivative and the algebraic interpretation of derivative.

We're going to do one more example of changes in a geometric form, and this example is an example that has to do with the Earth. So, suppose you took the entire Earth, and imagine the Earth to be literally a perfect sphere, of course it's not quite, but imagine it's a perfectly round sphere. Now, suppose we take the entire equator of the Earth, which is about 25,000 miles around, and we put a belt around the Earth—and you see this very attractive belt here in this graphic, just snuggly around the Earth. It's exactly along the Earth; and the Earth is perfectly round so it just touches it at every point. It just literally fits right there on that belt. It's exactly the length of the equator, about 25,000 miles. Now we say to ourselves, "Well, you know, I think that that belt is a little tight. The poor Earth needs a little bit of breathing room." And so, we decide to take the belt and loosen it up a little bit; take the belt and loosen it up. The way we'll loosen it up is we'll just, as you do with a belt, we'll just add a little bit more length to the belt. And just to be specific, let's add another 7 feet to the belt. So we make it—it's about 25,000 miles + 7 feet. You know, this much extra. Then we center it around the Earth. So, it's slightly off from the equator of the Earth, you follow me? So, it's just a little bit off.

Now, I'm going to ask you a question to see how your intuition is. How far off from the Earth do you think that belt would be? So, you've got this 25,000 miles around, and then you've added 7 feet, and how far off the Earth would it be. In other words, if this were part of the Earth, would the belt be just a very tiny bit off, or would it be this far off, or would it be this far off—all the way around? You've centered it so that it's literally centered around the Earth, and it's not touching. I'm going to give you a few seconds to think about this and make a guess, and let me ask you specifically, do you think you could crawl under it, for example?

Well, most people, their intuition says that if you've got something that's 25,000 miles around, and you add 7 feet, most people thing that that belt is going to be just barely off the ground. True answer? The belt is going to be a little more than a foot off the ground; and you can actually crawl under it, everywhere on the entire equator, it's more than a foot off the ground. How can we see that? Well, we an see it by thinking about what that increase of the circumference of the circle, which is adding distance to the belt, what does that correspond to? We're asking the question of the relationship between a change in the circumference and a change in the radius. Because, you see, the change in the radius is corresponding to how far off that belt will be from the Earth. That's what the change in the radius is.

So, let's look at this formula. Let's look at the formula that makes a relationship between the radius of a circle and its circumference. The formula is very simple. It is that if you have a circle of radius r, then its circumference is $2\pi r$. So, in the case of the Earth, you can think that the radius of the Earth is something like 4,000 miles, and so the circumference of the Earth is 2π times that radius; π is about 3, so this is a number that is a little more than 6, more than 6.25, but less than 7. That is, 2π is less than 7.

We saw that the derivative of this function is very simply computed since the graph of this function is a straight line with slope 2π; the derivative is simply 2π. Now, what does that mean? That means that for every 1 foot change in radius, it corresponds to a 2π, or about 6.28 and so on feet change in circumference. So, those two things correspond. In other words, if you increase the radius by a foot, you increase the circumference by 2 × π, which is about 6 feet. So, we can see that, in fact, increasing the circumference by more than 7 feet actually makes the radius have to increase by a rate—by more than 1 foot. So, this is something that often strikes the intuition a little aslant. So, that makes it fun.

Let's now turn our attention to another physical phenomenon, and this is the growth of a stalactite in a cave. Suppose we imagine that we're in a cave and we see stalactites growing from the ceiling. Now, a stalactite is—we're going to be viewing just conical stalactites; that is, stalactites that look exactly like cones. So, this is an example of the shape of the stalactite that we're considering. And as the stalactite grows, it adds material to it and it becomes a larger and larger cone. Now, that is to say, that's the model that we're going to be discussing in this next session. So, we're imagining that the cone grows so that it maintains the same shape; that is to say, the same ratio of its radius to its height. And, specifically, let's write down a formula for that. Let's say that for a stalactite of age x 1,000 years; let's suppose that it's $3x$ inches in height. So, stalactites don't grow very fast, by the way, so in 1,000 years we're imagining that it grows 3 inches. Okay? And then we're imagining that it maintains it shape; the shape being that the radius is always 1/6 of its height. So, it's sort of a pointy cone. That is, its radius is $x/2$—if you measure the years in thousands of years, then the radius is going to be half of that many inches in radius.

So, for example, at the 10,000-year mark of the stalactite, its height will be 30 inches and its radius will be 5 inches. At the 20,000-year mark its height will be 60 inches and its radius will be 10 inches. Now, the weight of a stalactite is proportional to how much volume there is in that stalactite. So, we can write down a formula for the volume of the stalactite, which we can interpret as its weight, if we use the correct value of weight per volume, it's proportional to the volume, so we'll just call this the weight of the stalactite, and we can compute it because we know that—in fact, we'll prove in a later lecture that the volume of a cone is equal to 1/3 the area of the base times the height. Don't just take my word for it for now, but later you won't have to take my word for it; we'll actually demonstrate

that that's the case. But, in this case, it's $\frac{1}{3} \pi \left(\frac{x}{2} \right)^2 (3x)$, which is

$\dfrac{\pi x^3}{4}$. That's the formula for the weight of the stalactite at every moment x where x is the number of thousands of years old it is. So, here we have it, and you can see that at 10,000 years old it's 250π pounds, and at 20,000 years it's $2,000\pi$ pounds in weight.

Now, here's a question for you. Do you imagine that the derivative of the weight function at 10,000 years is greater or smaller than the derivative of the weight function at 20,000 years? So, this is a question in interpreting the meaning of the derivative. It's talking about the rate of change of the weight of the stalactite with respect to time. Well, let's think about it. We know that the stalactite is growing at a constant height rate. But, then, material is being added to the outside as well. So, the volume increase for a unit change in the height, when the cone is smaller you're going to have a smaller layer of material, and therefore lighter, when you add 1 inch of thickness to that small stalactite. When you have a bigger stalactite, and you add 1 inch of thickness to it by adding 1 inch of length, then the material is greater. Therefore, what we're saying, the interpretation of it with respect to derivatives, is that the derivative of the weight at the longer time is going to be bigger than the derivative of the weight at the lesser age. So, this is an example, and we can actually compute it here using the formula for the derivative, we see that the derivative at the 10,000-year mark is 75π; that's the amount of additional weight per change in thousands of years; and at 20,000 years it's going to be 300π change in weight per change in year.

Okay. Let's do just one more quick example, and this is that we can also apply the analysis of looking at graphs and derivatives having to do with an economic situation. Suppose we have two graphs here, a supply and demand curve, where, depending on the price of an object, the suppliers are going to be inclined to produce more of them if the price goes higher. On the other hand, the demand curve is going to go down when the price goes higher. And, at some point, if these things cross, that is the place where you would expect the price to be, because if the price were higher than that, then more would be supplied than would be demanded, and so you wouldn't sell them all; and, if the price is less than that crossing point, then the supply is less than the demand, meaning that if the demand is bigger than the supply, then, if you increase the price, you still have more demand than supply, and so you could sell more at a higher price, and that would be an incentive for the people to supply more of these goods.

Well now let's look at how to interpret two different supply-demand graphs. If the supply and demand graphs cross sharply with sharp derivative, then this is an example of an elastic good, meaning that the price would tend to be very stable because any deviation from that price would lead to a greater imbalance. Whereas, if the supply and demand curves were almost tangent at the point of their crossing, then this is called an inelastic good, meaning that the amount of demand remains about the same for a large range of prices; things like how much gasoline you buy is of this character, because you have to have a certain amount of gasoline, regardless of the price of it, more or less; and so, there tends to be a big range in the variation of price of gasoline. So this is just an example of a case where by analyzing the slopes of functions and their relationship, we can deduce something about the world; in this case, in economics.

In the next lecture, we're going to turn our attention to some ancient developments that were precursors to the interval. I look forward to seeing you then.

Lecture Eight
Circles, Pyramids, Cones, and Spheres

Scope:

How do we find formulas for the areas of objects, such as circles, and the volumes of solids, such as cones, pyramids, and spheres? We can deduce each of these formulas by dividing the object into small pieces and seeing how the small pieces can be assembled to produce the whole. The area of a circle, πr^2, is a wonderful example of a formula that we may just remember with no real sense of why it's true. But we can view the circle in a way that shows clearly whence the formula arises. The process involves a neat method of breaking the circle into pieces and reassembling those pieces. This example and others illustrate techniques of computing areas and volumes that were ancient precursors to the modern idea of the integral.

Outline

I. Greek mathematicians had a keen sense of integral-like processes. We look first at an ancient process for discovering the formula for the area of a circle.

    A. Remember that the number π is the ratio of the circumference of a circle to its diameter.

    B. A circle of radius r can be broken into small wedges.

    C. The wedges can be assembled by alternately putting one up and one down to create a shape almost like a rectangle.

    D. As the wedges are made ever tinier, the assembled shape more and more closely approximates a rectangle.

    E. The top plus the bottom of the rectangle is precisely the circumference of the circle, therefore, of total length $2\pi r$.

    F. Thus, the top is πr long.

    G. The height of the rectangle is ever closer to r.

    H. The rectangle is approaching a rectangle of height r and width πr; hence, it has area $r\pi r$ or πr^2, the familiar formula for the area of a circle.

II. The derivative gives a dynamic view of the relationship between the area of a circle and its radius.

    A. We find that the derivative of the area (the change in area divided by change in radius) must equal the circumference of the circle.

    B. When we add thin bands to a circle to increase its size, then divide by the increment that we made to the radius, that division gives us the length of the circumference.

    C. Therefore, the derivative of πr^2 equals $2\pi r$. The derivative measures how fast the area of the circle is changing relative to a change in the radius.

III. The area of a triangle is dependent only on the height and base, not on whether or how much it is leaning.

    A. The area of a right triangle is easy to calculate because it is half of a rectangle.

    B. We can see that the area of any triangle depends only on its base and height by sliding thin pieces to the side to create a right triangle.

    C. Therefore, we find that the area of any triangle is equal to $\dfrac{1}{2}$ the base times the height.

    D. In fact, we can use this formula to calculate the area of a circle by imagining the circle divided into tiny triangles.

        1. All the triangles have the same height (r), and the sum of the bases of all triangles equals the circumference of the circle ($2\pi r$).

        2. Therefore, the sum of the area of all the triangles equals the total of the bases ($2\pi r$) times the height (r) divided by 2, or πr^2, again, the familiar formula for the area of a circle.

IV. We can determine the volume of a tetrahedron (a pyramid over a triangular base) by thinking of sliding its parts as we did with the triangle.

    A. The volume of a tetrahedron is determined only by the area of the base and its height, regardless of where the top point of the tetrahedron lies.

B. It is difficult to compute the volume of a tetrahedron, so we start with the volume of a prism.

C. The area of a prism is the area of its base times the height. Three tetrahedra of the same volume fill up a prism; thus, the volume of a tetrahedron is $\frac{1}{3}$ the area of its base times its height.

D. Once we know the volume of a tetrahedron, we can determine the volume of a pyramid.

    1. A pyramid can be seen as two tetrahedra if its square base is divided into two triangles.

    2. The volume of the pyramid, therefore, is equal to $\frac{1}{3}$ the area of half of its base times its height plus $\frac{1}{3}$ the area of half of its base times its height. In other words, the volume of a pyramid equals $\frac{1}{3}$ the area of the base times the height.

V. The volume of a cone is easy to compute once we know the volume of a tetrahedron.

    A. A cone can be approximately filled up by tetrahedra each with the same point as the cone point and with the bases of the tetrahedra on the base of the cone.

    B. Given that the volume of each tetrahedron is $\frac{1}{3}$ the area of its base times its height and the height is the same as the height of the cone, then the volume of the cone is also $\frac{1}{3}$ the area of the base of the cone times its height.

VI. The surface area of a sphere can be computed by breaking that surface into small pieces.

    A. The area between latitude lines on a sphere is the same as the area of a band around the surrounding cylinder if the band is contained between parallel planes that intersect the sphere on the two latitude lines. The smaller radii for latitudes near the

North Pole are accompanied by the "slantiness" at those higher latitudes in such a way that the area on the sphere between parallel planes near the North Pole exactly equals the area on the sphere between parallel planes of the same fixed distance apart near the equator.

B. We can see that this area equality holds by looking at a picture of a circle and finding two similar right triangles that tell the story.

C. Thus, the surface area of the sphere is exactly the same as the area of the surrounding cylinder, or $4\pi r^2$.

VII. These examples show ancient ideas that resemble the modern idea of the integral. The ancients did not have a well-defined idea of what happens at the limit, but their arguments are persuasive and can now be made mathematically rigorous.

Readings:

Boyer, Carl B. *The History of the Calculus and Its Conceptual Development*.

Dunham, William. *Journey through Genius: The Great Theorems of Mathematics*.

Questions to Consider:

1. In thinking about the surface area of the sphere, why couldn't we approximate the area of the northern hemisphere by just thinking of triangular wedges going from the North Pole down to the equator and viewing each wedge as a triangle?

2. Is there a philosophical reason that the formulas for areas and volumes are so relatively simple? In other words, could you imagine a geometric world in which the area of circle or the surface area of a sphere was not so simple? There is a whole different world of ideas associated with non-Euclidean geometries in which such formulas are not as simple.

Lecture Eight—Transcript
Circles, Pyramids, Cones, and Spheres

Welcome back to *Change in Motion: Calculus Made Clear*. In the last lecture we were talking about the derivative and seeing how the derivative is expressed in its relationship to the areas of objects and the volumes of objects and their formulas. But one thing that we didn't do last time is to say why the formula, for example, for the area of a circle is correct. So, in this lecture, we're going to ask the question how do we find the formulas for areas of objects, such as circles, and for things like the volumes of solids that are more complicated than a cube—whose volume we can easily see—but, how about the volumes for things such as cones, or pyramids, or spheres?

Well, it turns out that we can deduce each of those formulas by taking the object in question and dividing it up into many, many small pieces, and adding up those small pieces to get the formula for the whole assembled object. That concept, the concept of dividing things up into little pieces and adding them together, is the concept of the integral. So, what we're really going to be doing today is talking about the integral, an introduction to the integral, except that these are ideas that many of which occurred thousands of years before the integral was actually defined as we know it today. So, what we're talking about here is the concept that underlines the concept of the integral and how it appeared in deducing the formulas for areas and volumes. Let's get started.

Let's begin with the formula for the area of a circle. So, here we have a circle, and, of course, we all know the answer. The area for a circle is πr^2. Now, first if all, let's remember what π is. π is the number just above three—3.14159 and so on, just above 3; it's the ratio of the circumference of a circle to the diameter. In other words, the circumference is π times the diameter. Now, our question is: Why is it that a circle has area, π, that ratio, times the radius squared? Well, here's a very clever way to see that that formula is correct. Let's just take a circle and break it up into wedges. This, by the way, is important to do because that reminds us that π, if we think of the circle as a pie, then this is a good way to remember, breaking it up into pieces. So, we break it up into pieces, and here I've taken this circle and broken it up into eight pieces, and I'll just put two of them back here so you can see that, in fact, these are sectors of the circle, there are eight of them, and all I'm going to do is take them and rearrange them, put them cleverly one up, one down, one up, one down, and arrange them in this kind of a pattern. Well then we see that the area of these eight—one, two, three, four, five, six, seven, eight—these eight pieces, the area is, of course, the same as the area of a circle. And it's beginning to look a little bit like a rectangle.

Well, what happens if instead of dividing the circle into eight pieces, suppose that we divided it up into 16 pieces? Now we'll go to graphics so we can see this more effectively. Here's an example where we've divided the circle into 16 pie-shaped pieces, and put them alternately up and down, up and down, up and down. Now, notice something about this process. The circumference of the circle, which, remember is $2\pi r$ because 2 times r is the diameter and π times the diameter is the circumference of the circle, so we know that the total circumference of the circle is $2\pi r$, and yet that entire circumference, since half of the wedges are facing upward and half are facing down, we see that the total, sort of scalloped, top edge of this figure is exactly π times r in length; likewise, the bottom is also π times r in length; the totality being the circumference of the entire circle. And this slightly slanted length is r, the radius of the circle.

But now let's just think. What happens when we take this circle and divide it instead of into 16 pieces, into a million pieces, or a billion pieces, or a trillion pieces? When we do that, and then alternately put those pieces up-down, up-down, up-down, this becomes indistinguishable from just a straight line whose side is π times r in length; this will become a vertical line of length r; and the product or r times π of r, which is the area of a rectangle, is exactly equal to the area of a circle. So, this is a derivation of the formula for the area of a circle.

By the way, let me remind you, just as we did in the last lecture, that the derivative of the area does equal the circumference of the circle, and the reason for that is, as we discussed the last time, that when we add a small bit to the area of a circle, and then divide by the increment that we made to the radius, then that division is going to give us something that is, basically, just the length of the circumference. And, as you see, the derivative of πr^2 is equal to $2\pi r$. So, everything is fitting together.

Well, let's just move on and consider some other very familiar shapes and talk about how we're going to deduce the areas of these familiar shapes. Let's start with a very simple one, namely, a triangle. Suppose that we take a triangle—here, I'm going to move this picture of a triangle here. Now, if I take a triangle, and it's a right triangle, the area of a right triangle is very easy to compute, because if we put two right triangles together, we simply get a rectangle. And the area of a rectangle is very simple because it's just the base times the height. So, in that way, we see that for a right triangle, the area is equal to the base times the height, divided by 2. So, it's very simple to find the area of that kind of a triangle.

But, suppose that we have a triangle that is not a right triangle. How are we going to find its area? Well, here's what we're going to do. We're going to take our triangle and divide it up into little tiny pieces, and here we've given an indication of these horizontal slices that are making up our triangle. Now, notice something about a triangle and how it relates to other triangles in the following way. Suppose I take this triangle, and I simply slide the pieces over into a different configuration—see? When I slide the pieces over to this different configuration, notice that every single one of these horizontal strips simply slid over and is exactly the same length as it was before. And the reason for this is simple: that, if we have the same base to a triangle, and the same height, if we consider a band right in the middle, the area of that band is going to be exactly half—that is, the length of the band that's halfway up is going to be exactly half of the length of the base, because they're similar triangles. So, that means that when we do this sliding, we're going to get the same area as an area of a right triangle that had the same height and the same base.

Let's look at some graphics that illustrate this. We can slide the top of the peak of this triangle, slide it over to the side, and get sort of a slaunch-wise triangle, and notice that each of these horizontal bands just slides over neatly and fills up that slaunch-wise triangle. So, the area of a triangle is determined entirely by the length of its base and its height. It doesn't matter where that top point is in relation to the triangle. And, yet, we know that a right triangle has area 1/2 its base times its height, so we see that the area of every triangle is just the base times the height, divided by 2. By the way, it also gives us another way to see what the area of a circle is, because if we imagine a circle being made of tiny little triangles put together, we can see that the area of each triangle is its base times its height divided by 2. If we add up all of these triangles, we just get a bunch of triangles who have the same height, namely the radius r, and the sum of those bases is going to equal the circumference of the circle, $2\pi r$. So, the sum of all of those little tiny triangles would just equal the total of the bases, which would just be $2\pi r$, times the height, r, divided by 2, because each triangle has base times height over 2 as its area.

Now, let's take our philosophy of sliding things over and look at another kind of an object whose volume is a little bit trickier to figure out. Let's consider this object. This is a tetrahedron. It's like a pyramid over a triangular base. This is a tetrahedron. We'd like to find what the volume of this tetrahedron is, and the way we're going to analyze this is by realizing that a tetrahedron can be—although we don't know what it's volume is, we don't know a formula for its volume, that's what we intend to deduce in a minute. But one thing that we do know is that the volume of a tetrahedron that with a certain fixed base, as we have here, and a certain height, will be the same as a tetrahedron that has the same height and the same base, regardless of where the top point of the tetrahedron lies. Isn't this amazing? Because, you see, as I slide the top of the tetrahedron to different positions at the same height above the base—I'm trying to keep it at exactly the same height—you notice that all of these horizontal bands just slide and neatly fill up that tetrahedron. So, what that tells us is that the volume of a tetrahedron is determined by the area of its base and its height, and it's not determined by, for example, where that top point is. First of all, let me just say, I think this is a really neat kind of an illustration that shows when you move that top along all of those, like that middle one just slides and fills up exactly the correct position. It just finds its way.

So, our goal is to find the volume of a tetrahedron. Well this is a little bit tricky. The way we're going to do it is we're going to take an object that is a prism—now, this is an example of a prism. Its like—you've seen prisms that break light into colors. It's easy to compute the volume of a prism, because it's just the area of the base times the height. What's difficult is to determine the equation for the volume of a tetrahedron. But, what we're going to do is be very clever and see how a prism is made up of tetrahedra. So, here we have a picture of a prism, and we're going to take that prism, realize that its volume is the area of its base times the height—that's the volume of an entire of a prism, the area of the base times the height. If we take that prism, we're going to divide into three tetrahedra. Now, this is a little tricky to see, so you're going to have to focus for a second here. What I've done here in this illustration is to take our prism and illustrate that it's the union of three tetrahedra. One of the tetrahedra is this red one. The red one is obtained by taking the triangular top of the prism, and then taking the cone point down at this bottom vertex. So, you see that red thing is a tetrahedron. This is its base on top, and then the cone point is down here. Likewise, the white tetrahedron is constructed by considering the base of the prism, that's a triangle, and then taking this vertex at the top and coning up—remember a tetrahedron is like a pyramid over a triangular base. Well, the triangular base is the white base of the prism.

The tricky one to see is this blue one. This blue thing is, in fact, a tetrahedron, and the way to see it is this: the front part of the tetrahedron is a triangle—do you see this blue triangle? It's half of the rectangular side of this prism. So, this half is a triangle—that's a blue triangle—and then we cone down to this opposite point, and that creates a blue tetrahedron, and that blue tetrahedron creates the third tetrahedron that makes up the prism. The question is: Why do we think those three tetrahedra have the same volume? But, they do.

It's easy to convince ourselves that the volume of the white tetrahedron is equal to the volume of the red tetrahedron. The reason is that they both have identical sized bases, namely the top and the bottom of the prism, and then the height of the tetrahedron, the red one, is just the height of the prism and the height of the white tetrahedron is the height of the prism again. So, the white and the red tetrahedra are clearly the same volume. But now look at the white tetrahedron compared to the blue tetrahedron. The white tetrahedron can be viewed, instead of having its base at the bottom of the prism and then its cone point up at the top; instead, view it as the base of the tetrahedron being this side, this triangular half of this rectangular face of the prism, this white triangle, and then coning to this point. You see, a tetrahedron—you think of it as this being the cone point over a triangular base, but there's no reason we can't think of this as the base and then coning to this vertex. Any one of the sides of a tetrahedron can be viewed as the base and coning to the opposite vertex. In this way, we can see that this white tetrahedron can be viewed as having a base over here and this as the cone point, but then it's clear that its volume is the same as the blue tetrahedron, because, you see, the blue tetrahedron has an identical area base as this white one does, because they're just the two halves of a rectangle, and they also have exactly the same cone point. So, their height is also identical, therefore the volume of the blue is the same as the volume of the white, which is the same as the volume of the red tetrahedron. So, those three tetrahedra have the same volume as each other; and, consequently, the total volume of one tetrahedron is just one-third of the volume of the entire prism, which is the area of the base times the height. So, this has allowed us to deduce this formula for the volume of a tetrahedron. It's 1/3 the area of the base times the height.

Well, once we know the volume of a tetrahedron, we can also figure out what the volume of a pyramid is because, look, a pyramid can be written as just two tetrahedra. Now by the way, remember that a pyramid has a square base, and then it has a cone point over it, like the Great Pyramid at Giza. So, here is this great pyramid—you can divide it in half by simply taking the base and drawing a straight line down the middle, and then you can see that you're coning down to one of those triangles, making a tetrahedron half of it, and the other one is the other tetrahedron, another half. So that means that the volume of this pyramid is equal to 1/3 the height of the pyramid times the area of half of the base, plus 1/3 the height of the pyramid times the area of the other half of the base. By the distributive law, since both of those involve 1/3 times the height and we're just adding the areas of the base together, we see that the volume of a pyramid is just the area of the base times the height, divided by 3.

Well, that concept allows us to also compute the volume of a cone, because just think of a cone as being made up of many, many little triangles, and, of course, we might have to have infinitely many of them to fill up the base of the cone, but a cone, you see, has an area of the base—which could be thought of as adding a bunch of triangles together—and then it has a height, so the volume of the cone is equal to 1/3 the area of the base time the height, because it's just the sum of this formula for each of the tetrahedra that we can imagine to be making up that cone.

So, this process has allowed us to compute the volumes: formulas for the volume of a tetrahedron, for the volume of a pyramid, and the volume of a cone. Well, that's very nice. In fact, we can take these ideas and now we're going to try something that's a little bit trickier, perhaps, and that is to compute the surface area for the surface of a sphere. So, here we have a sphere and we imagine this sphere to have radius r, and our question now is going to be, what is the surface area of this sphere? Now, we can think of it as, for example, the Earth, and we're trying to find what is the surface area of the Earth. This is actually quite a tricky kind of concept because, of course, it's curved—it doesn't have a flat bottom with a cone or anything like that—we've got to figure out how to compute the surface area of the sphere.

Well, our strategy—as our strategy has been for all of these methods—is to break something up into small pieces and add those small pieces together. And that, by the way, is the philosophy of the integral, and that's why we're doing this lecture, because it's introducing the concept of the integral, which is to break things up into little pieces and add them together. So, let's go ahead and see if we can find a way to compute the area of the surface of a sphere.

The way we're going to think about it is this, and I'm going to pose a question to you: suppose that somebody were to make you an offer to buy some land on the Earth, and here's the way they made the offer, they said, "I'm going to take a slab of the Earth," and what I mean by that, I take a straight line from the North Pole to the South Pole, and I'm just going to cut it perpendicular to that line with a plane. So, that is a plane through a fixed latitude line—do you follow me? We're just going horizontally through the Earth in a plane parallel to the latitude. Now, we're going to take another latitude that's, say, a mile down. So, these are two parallel slices, go through two latitudes that are 1 mile vertical distance down, and they cut through the Earth, see, in a slab, and the boundary of the Earth, then, has this thin strip that goes all the way around. Now, here's a question for you: suppose somebody said to you, "I'm gong to give you an area of the Earth, and you can take it either right at the equator, where you have vertical distance of 1 mile, and it goes all the way around the equator, and all the area in that—you can take that, that is one of your options; or, alternatively, right about halfway in between here, the top and the bottom, we'll take this vertical distance of 1 mile, again, slice through and you could have that area, if you'd prefer. Or, even, right at the very top, you can take the area that goes from tangent to the North Pole down 1 mile, vertical distance 1 mile, and take the area of the Earth between those two bands. Which one would you choose to get the most area?"

Now, let me point out why this is sort of a challenge. At the equator, when you take this band that has this vertical distance 1 mile, the Earth is basically vertical at the equator. So, you're getting a strip that's sort of like a vertical belt. But, when you're halfway up in between the equator and the North Pole, the Earth is tilted somewhat, so although the circle at that latitude has a smaller radius, nevertheless, since it's tilted, the tilt gives more area. You see, because it's tilted, you have a longer strip along the Earth, and yet you have a smaller radius. So, these things have to be balanced. At the very top, you'd have just a little—like a northern polar cap—cap; it would be like a little cap—which one do you think you'd prefer, in the sense of just surface area? Of course, we're not talking about what's cold and what's warm, and you like the tropics—we're not talking about that; we just want to know which has the most area.

Well, the surprise is that, in fact, all of those options have exactly the same area. That is, if you take a sphere and you cut it by two parallel planes, the area, the surface area on that sphere between any two is going to be precisely the same, if the distance between the parallel planes is the same. Let's see why this is true.

Here's a picture of where we're imagining cutting our surface in this fashion, here between these two parallel planes, and I'm going to put a cylinder around the sphere, and the cylinder is just snuggly fits on the outside of the sphere—as you see—and the claim is going to be this: That, no matter where we cut the sphere, the area between the cuts on the sphere is precisely the same as the area of this band on the cylinder that's on the outside of the sphere.

I want to make absolutely sure you understand this, so let me draw this picture of what's going to happen when I cut right here between the sphere and, you see, I'm going to have a band here, and then I'm going to have this tilted blue part here; it's tilted because at that, you know in the northern hemisphere, the Earth, or a sphere, is going to be somewhat tilted. Now, we're going to compare those areas and we'll see that they're exactly the same. Now, here's how we're going to do it: imagine cutting the sphere by a vertical circle, and if we cut it by a vertical circle—so, let me make sure you're following this, we have this cylinder going around it like this, and now I'm going to cut with a vertical circle, right through this way, and it's going to cut through this outside cylinder in two vertical lines—a vertical line here and a vertical line here—and then it's going to cut the sphere in a circle, a great circle. Here we have it on the graphic—here are the vertical lines where it's cutting the cylinder and here is that vertical circle that cuts through the sphere. Now we're going to look at this diagonal piece of the sphere that's being cut, and then we have this vertical piece between these two vertical planes, here and here; we're going to see that this diagonal piece has a certain relationship to the other parts of the figure.

At this height, the radius of the sphere is s, that is if I go from this point right in the middle out to the outside; since it's near the top, its radius is smaller than r, and, in fact, we're going to call that s. This is the radius, s. The radius of the whole sphere, and therefore this circle, is r. The diagonal line, we're imagining we're taking such a small part of the sphere that the curve of the sphere is indistinguishable from a straight line, and we'll call its length Δs, and Δh is the vertical height. Now, this figure may look familiar to you because we actually saw it in a previous lecture, and notice that this triangle here is similar to this triangle up here, and I'll tell you why. They are both right triangles, and this angle here is equal to the opposite interior angle of a line cutting two parallel lines; so this angle right here is equal to this angle here; Δs is on a line that is tangent to the circle and therefore perpendicular to r; and, consequently, this angle is 90° minus this angle; and therefore this angle is exactly the same as the angle at the origin here.

In other words, if you didn't follow all those details, the fact is that this little triangle is similar to this bigger triangle; and, since they're similar triangles, we see that the hypotenuse of the small triangle, Δs, divided by Δh, the leg that is next to our angle that is the same as this angle, that that ratio is equal to the same thing as the hypotenuse of this bigger triangle divided by s, the radius at this height of the circle. Therefore, we have this little formula.

Well, once we have this formula, just a little tiny bit of algebra, taking this formula and cross-multiplying, multiplying both sides by 2π tells us that $2\pi r\,\Delta h$, which is equal to the area of this belt around the cylinder, is equal to $2\pi s\Delta s$, which is this diagonal length times the circumference of a circle of radius s. Consequently, as promised, the area on the sphere of that slanted slice is exactly equal to the vertical area on the boundary of this enclosing cylinder. Since that's true for every single slice, we can add them up and see that the area of the sphere is exactly equal to the area of this enclosing cylinder, but the enclosing cylinder has circumference $2\pi r$, it has height $2r$, so its area is $4\pi r^2$. So, the sphere, the surface area of the sphere, is also $4\pi r^2$ —four times the area of the circle of radius r.

Well, I think that all of the strategies for finding these equations for areas and volumes that we saw involved breaking up pieces of what it was we were trying to get the formula for, and breaking them up into little bits and adding them together. Those are ancient ideas that very closely resemble the modern idea of the integral. And the ancients had this wonderful concept that they used all the time, but they didn't define it in the clear way that we see as the limit of a process. But their arguments were very persuasive and lead to mathematically correct ideas that we can now make completely mathematically rigorous.

I look forward next time to showing you an amazing example of this kind of logic that Archimedes used to deduce the equation of the volume of a sphere. I'll see you then.

Lecture Nine
Archimedes and the Tractrix

Scope:

In the 17th century, Bonaventura Cavalieri analyzed shapes and found formulas for the areas and volumes of geometrical figures using his *method of indivisibles*. Previously, in the 3rd century B.C., Archimedes devised an ingenious method using levers to deduce the formula for the volume of a sphere. The method foreshadowed the idea of the integral in that it involved slicing the sphere into thin sections. The idea of the integral provides effective techniques for computing volumes of solids and areas of surfaces. Then we move to the 21st century and a new method for computing areas developed by mathematician Mamikon Mnatsakanian.

Outline

I. In the 17th century, Bonaventura Cavalieri analyzed shapes using his *method of indivisibles*.

   A. If one thinks of the surface of a sphere as comprised of many tiny triangles, then the volume of the sphere can be viewed as made up of many tiny tetrahedra with those triangles as bases and the center of the sphere as the top of each tetrahedron.

   B. Because we know that the volume of each tetrahedron is $\frac{1}{3}$ the area of the base times the height, then the volume of the sphere will be $\frac{1}{3}$ the surface area of the sphere times the radius (the height of each tetrahedron).

   C. Because the surface area of the sphere is $4\pi r^2$, as we saw in the last lecture, the volume of the sphere is $\frac{1}{3}$ of the product of that area times the height of each tetrahedron, which is r, thereby giving the formula for the volume of the sphere, namely, $\dfrac{4}{3}\pi r^3$.

II. Archimedes had an amazing way to discover the formula for the volume of a sphere of radius r.

    A. His method involved a lever.

    B. He balanced a cone (with a base of radius $2r$ and a height of $2r$) and a sphere (of radius r) on one side of the lever with a cylinder (with a radius of $2r$ and a height of $2r$) on the other.

    C. Archimedes's method for showing that the objects balance involved dividing the sphere, the cone, and the cylinder into thin slices and hanging those slices on the lever.

        1. Originally, one could picture the cylinder being in its horizontal position with the cone and sphere neatly inside it.

        2. Archimedes's insight was that if we take a thin slice through the cylinder (thereby cutting through the sphere and cone also), that thin slice of the cylinder (by itself, left where it is, at point x distance from the fulcrum) would be exactly counterbalanced if the slices of the cone and sphere were both moved to the other side of the lever at distance $2r$ from the fulcrum.

        3. Because that insight is true for each slice, the totality of all the slices all the way along the cylinder shows that the cylinder, cone, and sphere will balance on the lever as described above.

    D. In our demonstration, we can see that placing the sphere and cone exactly $2r$ from the fulcrum balances the cylinder.

        1. The cylinder lies horizontally along the lever with one end at the fulcrum and the other at $2r$.

        2. Both the cone and the sphere are hung from the same point on the other side of the lever, namely, at the point that is distance $2r$ from the fulcrum.

        3. If we know that the cone, sphere, and cylinder balance, and we know the volumes of all the objects except the sphere, then we can deduce the volume of the sphere.

E. Thus, Archimedes found that a sphere of radius r has a volume of $\frac{4}{3}\pi r^3$.

F. Today, we would formalize this procedure of slicing up a sphere by using integrals.

III. Let's examine the relationship between the surface area of a sphere and its volume.

A. Knowing the meaning of the derivative, we know that the derivative of the volume is telling us the rate at which the volume is changing relative to a change in the radius.

B. Geometrically, that rate of change is the volume of a thin layer over the surface of the sphere divided by the thickness. As the thickness gets tiny, that fraction will simply equal the surface area.

C. Using our knowledge of derivatives, we know algebraically that the derivative of $\frac{4}{3}\pi r^3$ is $4\pi r^2$, which is the formula for the surface area of a sphere.

D. The derivative of the volume of the sphere must give a formula for the surface area of the sphere—and we see it does.

IV. In the 21st century, Mamikon Mnatsakanian devised an ingenious method for computing areas by breaking up regions into pieces that are sectors of a circle.

A. The area between two concentric circles can be computed in two ways.

1. Just subtracting the area of the smaller circle from the area of the larger circle is one method to arrive at the area of the ring (annulus).

2. The Mnatsakanian method is to view the annulus as a polygon with many, many sides.

a. We can sweep and shift the small triangular segments that make up the annulus and see that the sum of those segments will be the area of a circle.

b. We find that the area of the annulus is πa^2, where a is the distance from a tangent point on the small circle to the outer circle.

B. This method also provides a proof of the Pythagorean Theorem.

C. This method can be used when computing the area under a tractrix.

    1. A *tractrix* is a curve created by pulling one end of a string along the x-axis while the other end, attached to a pen, starts on the y-axis and is dragged along to create the curve.

    2. One way to compute the area under the tractrix is to view that area as a sweeping of tangent lines and to approximate the area as segments of a circle. The method shows that the area is simply equal to a quarter of a circle $\left(\dfrac{1}{4} \pi a^2 \right)$.

    3. The hard way to compute the area under the curve involves finding the formula for a tractrix.

V. Both Archimedes's and Mnatsakanian's methods involve breaking up an object into small pieces and adding up their contributions. This strategy is the fundamental strategy of the integral.

Readings:

Boyer, Carl B. *The History of the Calculus and Its Conceptual Development.*

Dunham, William. *Journey through Genius: The Great Theorems of Mathematics.*

Mnatsakanian, Mamikon. *Visual Calculus by Mamikon.* www.its.caltech.edu/~mamikon/calculus.html.

Questions to Consider:

1. Experiment with a level to see that the formula for balancing on a lever is correct. See how putting two weights on one side is equivalent to putting one weight at a different location. What location?

2. Deduce the formula for the area of a circle using Cavalieri's method of indivisibles.

Lecture Nine—Transcript
Archimedes and the Tractrix

Welcome back. Remember that in the last lecture we began a process of seeing how we could find the formulas of the areas and volumes of familiar figures using techniques that involved chopping up those figures into little piece and adding them together. Today we're going to look at a couple of examples of this strategy that preceded the definition of the integral, and then some examples that actually are very modern.

We'll begin in the 17th century, when a mathematician by the name of Bonaventura Cavalieri analyzed shapes and found the formulas for the areas and volumes of various geometrical figures using a method that was called his "method of indivisibles." So, I wanted to just tell you how he deduced the equation for the volume of a sphere using the kind of knowledge that we just figured out from the last lecture.

Remember, in the last lecture, we deduced what the equation was for the volume of a tetrahedron. Remember that the volume of a tetrahedron turned out to be the area of the base times the height divided by 3; 1/3 the product of the area of the base times the height. Now, what Cavalieri did was to say let's imagine a sphere as composed of a collection of tetrahedra, basically, where the base of the tetrahedron is on the surface of the sphere, and they just cone down to the center of the sphere. So, here in this graphic, we can see an example of one of these little imagined tetrahedra that cone down to the center of the sphere. Well, if we imagine covering the entire surface of the sphere with little tiny tetrahedra, so that the fact that the tetrahedras's bases are flat doesn't make any significant difference, the total of the volume of each tetrahedron is the area of its base times the height divided by 3, but since we're talking about summing over the sum of the bases of the tetrahedron being equal to the entire surface of the sphere, then we can see that the volume of the sphere will just be the surface area of the sphere times the radius, which is the height of each of those tetrahedra that we're thinking of putting together to compose this total solid sphere, divided by 3; that is, the surface area of the sphere, which last time we computed to be $4\pi r^2$, so that's the total surface area of the sphere—and think of it as the sum of the areas of the bases of all of these tetrahedra that are filling up the sphere—times the height of each tetrahedron, which is just r, each one is r, then times 1/3. So, the volume that we get for the equation that Cavalieri's method gives us for the volume of the sphere is then just $1/3\,r$ times the area of the sphere, $4\pi r^2$. Combining them gives the correct formula that the volume of a sphere is $\dfrac{4}{3}\pi r^3$.

So, that's one method for finding the formula for the volume of the sphere, but the one that I think is really amazing is a method that Archimedes devised. Archimedes was a mathematician from Syracuse, who lived in the 3rd century B.C., and he was just amazing and did just all sorts of incredible things, but this one may take the prize, because in this one he used his method of levers to find the formula for the volume of a sphere. You may remember the way levers work—this is the way a lever works: If you have a lever, for example a teeter-totter, and you have a heavy person on the teeter-totter and a little child on the teeter-totter, in order to get the teeter-totter to balance, the heavy person stays closer to the fulcrum and the lighter person goes further out, because a teeter-totter will balance if the product of the distance times the weight on one side is equal to the distance times the weight on the other. That's the principle that Archimedes used in computing the volume of a sphere, and let me show you how he did this.

The story starts with three objects. One is the sphere, whose volume we wish to find a formula for. So, this is a sphere. This is a cone. Now, this cone has exactly the same height as the diameter of the sphere; that is $2r$. So, this height right here, from the top of the cone down to the base is $2r$, and notice that this is exactly the same height. Then, the base of this cone, by the way, is exactly $2r$ in radius; in other words, the diameter of this is equal to the radius of the base of this cone. Likewise, this is a third object. This object is a solid cylinder. It has radius, once again, $2r$; the radius is $2 \times r$, that is the diameter of the sphere, and you can see that the height of this cylinder is exactly the same as the height of the sphere, and it's exactly the same as the height of this cone.

Okay, why do we have these three things? It's not at all clear. So, let me explain how Archimedes thought about this. He imagined putting all three of these things in the same location. Now, of course, you can't actually do this, but we're imagining this cone as being inside of this 3-dimensional solid, and this sphere as being right in there as well, and arranging them in this following way, which we can visualize, even though, of course, we can't literally put them—because they're solid we can't put them in the same place, but we can visualize them as being in the same place—and this is a picture of this big cylinder here, with radius $2r$ enclosing the sphere, whose diameter is $2r$, and enclosing this cone, here having base the same as the base of the cylinder, and coning to the fulcrum point.

Now, of course, if we put all three of these objects on one side of this lever, it's obviously not going to balance. So, this is where Archimedes starts to analyze how to balance pieces. The way he thinks is the following: Let's just chop this entire picture by a plane, a very thin slice that we take that is some distance away from the fulcrum. If we do that, it slices through all three of our objects; it slices through the big cylinder, and it creates a disk whose radius is $2r$. It slices through part of the cone; and it slices through a part of the sphere. Here's a picture of those three things; they're just concentric disks inside one another. This is what Archimedes observed. He observed that if you take the disks that come from the sphere and from the cone—the disks that were exactly at this place and move them over to exactly $2r$ distance on the other side of the fulcrum—so he hangs a disk over exactly point $2r$, and he hangs another disk, the disk from the cone, from the cone and the sphere, they are hanging at distance $2r$ from the fulcrum, and then he just leaves the big disk from the cylinder exactly where it is, at this point x distance away from the fulcrum. The fact is that those objects balance one another. They will stay in exact balance if they're hung from this fulcrum.

The reason that they hang from the exact balance, we can actually compute this using this diagram. This is a sideways slice of seeing our sphere here, the cone coming out, and then the big cylinder is not pictured. Using this diagram, we can compute what the radius is of the part of the sphere that is distance x away from the fulcrum. And the answer is its radius is $\sqrt{2xr - x^2}$ and the radius of the cone—the radius of the disk that comes from the cone—well, since this distance is x away from the fulcrum, and the cone has the property that when it goes $2r$ distance this way, it goes $2r$ distance up, then that means that if we're x distance away from the fulcrum, then the radius of the cone at that point is also x.

Using that formula and the Pythagorean Theorem, we can actually deduce that, in fact, those disks put at distance $2r$ from the fulcrum will exactly balance a disk of radius $2r$ that's left alone at that place x. Let's not worry about the details, but realize the implications. The implication is this: That if we do this balancing of the part of the cylinder at every point from the fulcrum out to $2r$, and we accumulate all of the disks that we get from the sphere and all of the disks that we get from the cylinder, we'll have the entire sphere here, the cone, all hanging from distance $2r$ away from the fulcrum, and then the big cylinder is just going to be fixed where it began, and the claim is that those two things will balance.

Now, what I'm going to do now is demonstrate that they balance by actually, physically doing it. Here we have the cylinder, the sphere, and the cone, and I'm going to attempt to balance them. So, I'll take them over here. So, on one side of this lever—here we have this lever; it's a balanced object here—on one side we have the entire cylinder sitting exactly where it began. Now, balancing the sphere and the cone exactly distance $2r$, that is, the diameter distance away from the fulcrum, exactly balances this cylinder. Well, first of all, it's, I think, rather impressive that this actually works. So, that is something to celebrate. I'm impressed that this is balancing, but what it does is it allows us to take our knowledge of what the volume equation is for a cone, and what the equation is for the volume of a cylinder, and the fact that they balance will allow us to deduce what it is that is the formula for the volume of a sphere. Let's go ahead and do the computation. Here it is:

We know that these two objects, the sphere and the cone, are both hanging at a distance $2r$ away from the fulcrum. Now remember, in order for something to balance, the distance times the weight on one side has to equal the distance times the weight on the other. Well, the distance away is $2r$, and then the weight is the weight of the sphere plus the weight of the cone. On this side of the equation, we have this cylinder balanced. Now, of course, it's spread out between zero distance all the way up to $2r$ distance, but that's the same as having all of its weight exactly distance r away from the fulcrum. So, its distance from the fulcrum is r times the weight of the cylinder and these two things balance because they balance on this lever. So, we know the formulas for the cone—it's 1/3 the area of the base times the height; we know the formula for the cylinder—it's just the area of the base, $\pi \times 2r^2 \times$ its height, and it's distance r away, and this side is $2r$ away. So, these things balance. Now we just do a little bit of algebra, and going through there we solve for a sphere, which is the volume of a sphere, and we see that the volume of a sphere is, indeed, $\dfrac{4}{3} \pi r^3$. Now, this is an amazing way to deduce the formula for the volume of a sphere. So, Archimedes was an amazing, amazing mathematician, and this is just one of many wonderful things that he accomplished.

By they way, let's just notice something about the equation for the volume of the sphere, that if we imagine taking the derivative of the volume of the sphere—remember, the formula for the derivative is just taking the volume of a radius plus a Δr minus the volume of the sphere over Δr, we see that, on the one hand, we know that when we divide this incremental increase in volume by this incremental change in the radius Δr, we just get something that's a thickening up of the surface of that sphere divided by Δr; and, on the other hand, we get the fact that we can take the derivative using the formulaic method that we saw in the previous lecture, we saw that its derivative, the derivative of $\dfrac{4}{3} \pi r^3$, bring down the 3, reduce the exponent, we get $4\pi r^2$, which, remember, was the formula for the surface area of the sphere. So, once again, we see that the geometry fits together with the beautiful mathematics.

Now, what we're going to do is take a step into modern times, and this is actually rather amazing to me, that I'm now going to be talking about some work that was published in papers that were written in the 21st century—that is just a few years ago—by a mathematician by the name of Mamikon Mnatsakanian and Tom Apostle. The two of them wrote up this work, and these ideas are owing to Mamikon Mnatsakanian, and these are beautiful concepts. Of course, these did not precede the definition of an integral, but they very well may have been ideas that Archimedes himself had thought of. But, of course, much of Archimedes's work was lost, and Mnatsakanian himself has said that maybe Archimedes actually did think of these ideas. But, I think that you'll find them very intriguing. Once again, they illustrate the concept of breaking up regions into small pieces and fitting them together as a method for computing areas.

So, here's what Mnatsakanian observed. He said suppose that you have an object, for example an octagon. Now, here I've drawn an octagon; you can probably see this eight-sided figure here, and Mnatsakanian said the following: suppose we do the following procedure. That is, we simply take a sector of a circle and extend one of the legs of the octagon and just sweep down until we are touching that leg of the octagon. Then just shift if forward so that we can just sweep some more, sweep some more, shift it forward so we can sweep some more, and so on, all the way around the octagon. Now, if we do this, that is, we always are taking the same radial distance and sweeping it down, each of these pieces is a sector of a circle. The totality of all of these sectors we get, since the sum total of everything that we're doing is to sweep all the way around a circle; that is, we sweep, and then shift over; sweep, shift over; sweep, shift over; sweep, shift over; and so on, that all of these pieces put together will simply be a circle. That's pretty clear. And, by the way, it doesn't matter that it's an octagon; it could be a square, it could be a triangle, it could be a 100-sided figure, anything that we do as long as we take a straight line and then just sweep it and go to the next corner and sweep it and go to the next corner and sweep it—once we go all the way around we will, of course, have swept a totality equal to a circle whose radius is the distance of the sector that we swept.

This is interesting and a rather simple observation about things. Let's go ahead and on the graphics see that we can do it with a 16-sided figure. Something to notice, by the way, is that if we take a larger figure, for example a larger octagon, but kept the sectors of the circle exactly the same radius, then the sum of those sectors around them will still be the same total area, the same area of the circle. Now, so far these are rather simple observations. We can do the same thing with a small 16-sided figure and a larger 16-sided figure. So far those observations may not be all that interesting, but they have some very interesting consequences. So, let's look at the following:

Suppose we ask the following question: What is the area between two concentric circles? Well, of course, the natural thing to do is just to take the area of the outer concentric circle and subtract the area of the inner concentric circle, and then the difference would be the area of that ring. By the way, that ring-shaped thing is called an annulus.

So, that's certainly one way to find it—just take the area of the outer circle minus the area of the inner circle—but let's think about it in a different way, and let's think about it using the Mnatsakanian strategy. You see, his strategy was we'll take this circle, and instead of thinking of it as a round circle, the inner inside circle, let's approximate it by a polygon with many, many different sides. So, for example, we could start with our 16-sided polygon, and then take our radius and put together those segments just as we did before, and we would see that the sum of those segments shifting and sweeping, shifting and sweeping, shifting and sweeping, we know that the sum of those segments, just as we saw in this original picture, will just be the same as a circle; the total area will be the area of a circle whose radius is the distance that we swept. Now, if we approximated this inner circle by a polygon that had, for example, a million sides—a million-sided polygon; in other words, it would be indistinguishable from an actual circle, look what would happen. We would have a little tiny sliver, a little sector, and then another little sector, another little sector, and sweeping those sectors around the totality of those sectors would be indistinguishable from this ring. Yet, the insight of Mnatsakanian is that the sum of those sectors could just be recombined to be just a regular circle. So, the area of this ring is just $\pi \times a^2$, where a is the distance from a tangent point on the small circle out until it hits the outer circle.

Now, this is rather interesting because look, suppose that we think about having a bigger inner circle, and we ask the question: How big should a larger circle be so that the ring between those two concentric circles would have exactly the same area as the ring over here? Well, we know the answer, because if we use his exact same method, and we start with a larger circle that is hit at distance a away from the tangent line coming out from the inner circle, we know that we could sweep that line around, or, if you prefer, we could approximate this inner circle by a polygon of a million sides and think about putting together all these little sectors, and those would combine to be a circle whose radius is a. This is rather interesting because what it tells us is that if we want two rings to have the same area—it's a very simple thing to do—all we do is we make the distance from the top point tangentially off until it hits the outer ring to be distance a, and then those rings will have exactly the same area.

Well, that's rather interesting. It's interesting in its own right, but also it's surprising that it actually provides a proof of the

Pythagorean Theorem. Now, let me demonstrate this. You see, because what we showed was that if we have a big circle and a small circle, and we wished to know what the area of that ring is, we saw that the area of the ring was just equal to the area of a circle whose radius is this distance a; but, but on the other hand, we also know the original method for finding the area between the area of a ring, the original method that we thought about was to take the area of the outer circle minus the area of the inner circle. Well, let's just do that. The area of the outer circle—let's suppose the radius of the outer circle is R, then the area of the outer circle is πR^2; and if we subtract the area of the inner circle, the inner circle is πr^2; we will get πa^2. That is, we already know from Mnatsakanian's analysis what the area of that ring is—it's πa^2. So, we have this formula, using the analysis, just factoring out—canceling out—the π's, we see that $R^2 = a^2 + r^2$, which is the Pythagorean Theorem. So, we actually can deduce the Pythagorean Theorem from Mnatsakanian's analysis of the area of that ring.

Well, this is really, I think, a very clever concept, and I wanted to show you another example of how this kind of insight can be used to find an area that is very complicated to find using other methods. So, let's clear this off for a minute. What we're going to be talking about now is trying to find the area underneath a curve that's a little bit complicated to draw. So, let's begin by drawing this curve. Now, this curve that I'm about to draw has a name, it's called a tractrix; and I'll first describe what I'm going to do, and then do it. A tractrix is a curve that's obtained by the following thing: We take a point that's vertically above—thinking about it on the vertical axis, the y-axis; it's some fixed distance away from the origin, and what I'm going to do is just drag my hand to the right along the x-axis, and it will cause this pen to be pulled along. The word "tractrix," think of it as tractor, and so it's a way of pulling, and it creates a curve. Here we go, when we do this, I'm just pulling along the x-axis, and you can see it's drawing a curve just by being pulled along. This curve becomes closer and closer to horizontal and it continues to get closer and closer, and eventually gets very, very close to the axis.

Now, it was a very difficult problem for people to write down an equation for this curve, and then to find the area under the curve was a very complicated thing to do; that involved calculus and it can be done with calculus. But I want to point out something about it that allows Mnatsakanian's method to allow us to find the area under this infinite curve in a very clever way. What I'm going to observe is at any point of this, as it's being pulled, notice that this line segment that is pulling it is always tangent to the curve that it's drawing; and the reason that it's tangent to the curve that it's drawing is this: that when I'm pulling to the right on the x-axis, I could resolve the forces of how the forces are acting on this point where the pen is, by looking at one force that goes directly in the direction of the segment, and then a perpendicular force of sufficient size, so that the two of them together equal the force that I'm actually adding along the x-axis. When we have that kind of resolution of forces, we notice that the vertical force, the perpendicular one, has no effect on this curve because it's like a circle, it doesn't have any pulling effect. So that means that the pulling effect is exactly straight along this, the direction of the line between the point and the x-axis, and consequently the line is tangent to the curve that it's drawing.

Well, here's what Mnatsakanian's strategy is: His strategy is to say instead of thinking of a smooth curve, suppose I think of it as a bumpy curve. That is, I go straight for a while, straight for a while, straight for a while, straight for a while; and then I draw a sector of a circle where the radius of the circle is the distance between the point and the point that I'm pulling; that is, the pen and the point I'm pulling, that distance a; that's the original height here, this distance is a, and then I can approximate the distance underneath the tractrix as a sector, another sector of the circle, another sector of the circle, another sector of the circle; and, if I divide my tractrix into more and more smaller and smaller pieces, the sum of those segments can all be combined back together again to just form a quarter of a circle, you see? We swish a certain amount, we make a sector of a certain angle, and then we start there and we swish another certain amount, and we start there and swish a certain amount, and so on. As they get smaller and smaller, those become indistinguishable from just starting—they can be moved back to be starting at the same radial point, and just construct a quarter of a circle, because it becomes almost flat at the end. So, the total area under the curve is just a quarter of a circle, $\dfrac{1}{4} \pi a^2$.

Well, you'd be very much impressed with this analysis if you had first worked on trying to write down the formula for a tractrix, which involves hyperbolic trigonometric functions. In any case, I thought you would be amused to see all of these methods that are really examples of the philosophy of how the integral computes things— breaking things into small pieces and adding up their contributions to get their total.

I look forward to seeing you next time.

Lecture Ten
The Integral and the Fundamental Theorem

Scope:

If a car goes at a constant velocity of 30 miles per hour, it is a simple matter to compute how far the car has traveled during an interval of time. We saw that to deal with varying velocity, we just break the total time into small intervals and add up approximations of how far the car traveled in each small interval of time. In this lecture, we will see the geometric implications of this integral process as we view it in graphical form. In particular, we see that the same process that computes the distance traveled by the car also computes the area between the graph of the velocity curve and the axis. We use Leibniz's notation for the integral because the long S shape reminds us that the definition of the integral involves sums.

Outline

I. After our introduction to the precursors to the modern concept of the integral, we will start a series of lectures that correspond to previous lectures about the derivative.

   A. We will see, first, the integral in its graphical interpretation.

   B. We will then study the integral in its algebraic interpretation.

II. Recall how the integral was defined in the case of the car moving down a straight road.

   A. We are given the velocity function $v(t)$ and want to compute the total distance traveled.

   B. For example, if we know the car was traveling a constant 2 miles per minute for 3 minutes, we know the car traveled a total distance of 6 miles.

III. Let's look graphically at the scenario of a forward-moving car.

   A. Notice that the process of finding the distance traveled involves finding products that are equal to the areas of rectangles.

   B. That is, the distance is equal to the product of the height of the rectangle (the line representing a constant 2 miles per minute) times the width of the rectangle (the 3-minute mark on the horizontal axis) of the graph.

IV. Let's look at another velocity function in which velocity is two times the time, or $2t$, so we know the car will be traveling along at an ever-increasing velocity.

   A. Our graph in this case shows an upwardly sloping line. To compute the distance traveled, we break the interval of time into small bits and do some adding.

   B. We then approximate the velocity traveled within those small intervals of time (half-minute intervals) by assuming our car remained at a constant velocity during that time, then "jumped" to the next constant velocity at the next interval and so forth.

   C. We then add the distance for the first interval of time plus the distance for the second interval of time and so forth.

   D. Notice again that the process of finding the distance traveled involves finding products that are equal to the areas of rectangles.

   E. As the little intervals get smaller, the total of the area of the thin rectangles is getting ever closer to the area between the curve and the axis; that is, the approximations improve.

   F. This infinite process of taking ever-smaller intervals of time provides us with a single exact answer.

V. Let us add here that the notation for the integral is Leibniz's, namely, $\int_a^b v(t)dt$.

   A. The long S shape reminds us that the meaning of the integral involves *sums*.

   B. The a and b denote the starting and ending times, respectively.

   C. In the limit, the answer is exactly equal to the area under the curve.

   D. Thus, in integral notation form, $\int_a^b v(t)dt$, which we know is the distance traveled by the moving car, is also equal to the area under the graph of $v(t)$.

VI. Let's look at the specific example where the velocity at each moment is $2t$.

    A. Then, $\int_0^3 2t\,dt$ is equal to the distance the car traveled between time 0 and time 3.

    B. But $\int_0^3 2t\,dt$ is also equal to the area under the graph of $2t$.

    C. We can check that area geometrically because the area under $2t$ between $t = 0$ and $t = 3$ is just a triangle with base 3 and height 6.

    D. Thus, the car traveled $\dfrac{1}{2}(6 \times 3) = 9$ miles.

    E. And the area under the graph of $2t$ from 0 to 3 is 9 square units.

VII. We can think about the motion of the car to see some features of the integral.

    A. The integral from a to b plus the integral from b to c equals the integral from a to c.

    B. This is obvious because it simply says that we see how far we went during the time a to b and how far we went from time b to c; the total is how far we went from time a to time c.

    C. Suppose the velocity is negative.

        1. When the velocity is negative, we are traveling backward.

        2. Then, the integral is telling how far backward we traveled.

    D. More exactly, the integral of the velocity is telling us not how far we drove, but how far we end up from where we started.

    E. Examples of when we are going forward part of the time and backward part of the time illustrate this concept.

VIII. Let's look at the graphical interpretation of integrals again.

    A. If the function is below the axis, then the integral is negative.

    B. If the function is part above the axis and part below, the integral combines the two.

C. The definition of the integral just adds up the products, not of the height of the rectangles, but the signed height—positive if the function is positive and negative if the function is negative.

D. It's easy. When the graph goes below the axis, the integral is negative; when above, positive.

E. The summation fact, a to b plus b to c equals a to c, works regardless of whether the graph goes above or below the axis.

IX. Integrals behave "opposite" of derivatives graphically.

    A. For a function $f(x)$, we can define a function $F(x)$ as the integral of f from starting time a to ending time x. Think of f as the velocity and F as the mileage marker.

    B. Recall that if a function $f(x)$ is increasing, then its derivative $f'(x)$ is positive, and if a function $f(x)$ is decreasing, then its derivative $f'(x)$ is negative.

    C. For the integral, it's the opposite: If the function $f(x)$ is positive, then its integral $F(x)$ is increasing; if the function $f(x)$ is negative, then its integral $F(x)$ is decreasing.

    D. If the function $f(x)$ is zero, then its integral $F(x)$ is constant—it's just $F(a)$ because we are not adding any area.

    E. In summary, notice that we have $F'(x) = f(x)$.

X. Now let's turn to an algebraic representation of this same idea.

    A. Given the equation for the velocity of a body, we can deduce the equation for its position using the integral.

    B. The integral of the velocity is the position.

        1. Let's look at an example where we know the answer.

        2. Suppose $v(t) = 2t$. We do some calculating and see that the distance traveled is the height times the width divided by 2, or t^2.

        3. If we stop at any time, the integral will give an answer.

        4. The answer is always t^2.

XI. Recall that the Fundamental Theorem of Calculus relates the integral and the derivative.

A. If $v(t)$ is the velocity at every moment of a car moving down a straight road, then the integral of $v(t)$ between one time and another equals the net distance traveled.

B. Thus, $\int_a^b v(t)dt$ equals the net distance traveled between time a and time b.

C. We saw before that if we can find a position function $p(t)$ whose derivative is $v(t)$, then the integral is easy to compute.

    1. It is simply the position at time b minus the position at time a.

    2. If $p'(t) = v(t)$, then $\int_a^b v(t)dt = p(b) - p(a)$.

D. Whether the variable is t or any other letter, the relationship is the same.

E. Thus, the Fundamental Theorem of Calculus states: $\int_a^b F'(x)dx = F(b) - F(a)$.

F. Suppose we want to do the integral process (which involves doing infinitely many approximations, each of which is a sum) for some function $f(x)$. If we can find an *antiderivative* $G(x)$ for $f(x)$ (that is, a function such that $G'(x) = f(x)$), then we can get the answer to the integral process just by doing one subtraction, $G(b) - G(a)$. An *antiderivative* is a function whose derivative is the function whose integral we are trying to take.

G. When we are faced with an integral, our first thought is, "Can we find an antiderivative for the function that is under the integral sign?"

XII. Let's look at some examples.

A. Every derivative formula leads to an antiderivative formula, because we can just go backward.

B. For example, we know that the derivative of x^2 is $2x$. Thus, an antiderivative of $2x$ is x^2.

C. The derivative of a constant times x, $f(x) = cx$, is just c. That is, $f'(x) = c$, or $\frac{d}{dx}(cx) = c$. Thus, an antiderivative of the constant function c is just cx.

D. The derivative of $\frac{x^{n+1}}{n+1}$ is x^n. Thus, an antiderivative of x^n is $\frac{x^{n+1}}{n+1}$.

XIII. Why do we say *an* antiderivative rather than *the* antiderivative?

   A. Any two functions that differ by a constant value will automatically have the same derivative at each point.

   B. We can see this fact graphically.

      1. If two functions differ by a constant, then their graphs are merely shifted up and down.

      2. The slope of the tangent line above each point will be precisely the same.

      3. The derivative is just measuring the slope of the tangent line.

   C. For each function, instead of just one antiderivative, we really find one antiderivative, then add any constant to indicate that any shifting of the antiderivative is also an antiderivative of the same function.

   D. Just as we saw the table of derivatives in Lecture Six, now we have the table of antiderivatives.

Function	Antiderivative
$f(x)$	$F(x)$
x	$\dfrac{x^2}{2} + C$
x^2	$\dfrac{x^3}{3} + C$
x^3	$\dfrac{x^4}{4} + C$

x^n	$\dfrac{x^{n+1}}{n+1} + C$
$\sin x$	$-\cos x + C$
$\cos x$	$\sin x + C$
e^x	$e^x + C$

XIV. The most important thing to remember about integrals is what they mean.

   A. The integral is a number that is the result of doing an infinite process.

   B. The process involves approximating the answer.

   C. Each approximation is obtained by taking a sum.

   D. As we divide the interval up into smaller subintervals, the approximation gets better.

   E. For positive functions, the value of the integral is equal to the area under the curve and above the axis.

Readings:

Any standard calculus textbook, section defining the definite integral and exploring its properties.

Questions to Consider:

1. The integral is defined in terms of sums of products. One number in the product is the value of the function in a small interval. Why does it not matter which point in the interval you choose in defining the integral?

2. Explain why an integral that gives the volume of a solid is somehow adding up pieces of area that have no volume and combining them to create volume. That almost paradoxical perspective was an obstacle to understanding the integral for many years.

3. Use the Fundamental Theorem of Calculus to evaluate $\int_1^2 x^3 \, dx$.

Lecture Ten—Transcript
The Integral and the Fundamental Theorem

Welcome back. During the last two lectures, we saw some examples of ancient work that were precursors to the modern concept of the integral. In this lecture we're going to start a series of lectures that correspond to the previous lectures that we had about the derivative. As you recall, we explored the derivative's manifestations in various places. We saw its relationship to graphs; and then we saw an algebraic manifestation of the derivative; and then applications of the derivative, interpretations of the derivative in many application settings.

In this lecture we're going to do the first two of those manifestations of the integral; namely, the integral and its graphical interpretation and its algebraic interpretation. So, to begin, let's begin by recalling the definition of the integral, and, in particular, remember that it was generated—the concept of the integral—was generated by our viewing the car that was moving down a straight road where we were looking at the velocity of the car at every moment, and then computing from that information the total distance traveled, the net distance traveled, by the car. And, so, here we go. If, for example, the velocity was at a constant 2 miles per minute, then we would discover that after 3 minutes if we wanted to know how far the car had traveled in 3 minutes, we would simply multiply the 2 miles per minute × the 3 minutes to get a total of 6 miles. And, by the way, I'm going to use this notation and talk a little about it in just a minute; this notation of the integral of $v(t)$.

Notice that that computation, the computation that gave us the total distance traveled, has another interpretation in this graph; namely, the interpretation is that it's the area between the graph of this function, just this very simple constant function, 2 miles per minute, the horizontal line, it's the area underneath that graph and above the horizontal axis, the time axis. Because the height is 2, that's the velocity; and the width is 3, that's the number of minutes that passed. And so 2×3 is the area under the curve. So, we see a correspondence between the computation that computed the distance traveled, velocity × time, and a computation that tells us an area, the area underneath the graph that we're taking the integral of. We'll see that this same procedure occurs for the integral in general. Suppose we have another velocity function; velocity at every moment is 2 × the time; so, our car is constantly speeding up as time goes from 0 forward. If we draw a graph of that velocity function, it's just this diagonal line; it's a straight line with slope 2. And, if we wish to compute the distance traveled, recall we had a strategy for that. Our strategy was that we broke the interval between 0 and 3, if we're trying to figure out how far we went during those first 3 minutes, we broke that interval into small increments of time, in this example we've broken it into 1/2-minute intervals, and then we approximated the speed that the car traveled in each of those sub-intervals of time. We approximated it by assuming that the car just stayed at a constant speed during each of those sub-intervals, and then jumped to the next constant speed, and the next constant speed, and so on throughout the interval.

So, we have this characteristic sum, where we have a product of the velocity at a steady velocity for the first interval of time plus a steady velocity for the second interval of time. We multiplied the two to get a distance traveled. If the car had been going at a steady speed for this interval of time, the distance traveled during that 1/2 minute of time would be the rate of speed, that is, the velocity at that time times how long we assume that the car went at that time.—in this case, 2 miles per minute × this 1/2 minute. That's the distance traveled if the car were traveling at that constant speed.

Now, I want to point out two things about this. First of all, notice that the procedure by which we are deducing the distance traveled by this approximating vehicle, that the distance traveled is also computing something else—it's computing the area of each of these rectangles, because each of these products is the product of the height of—that is, the value of the velocity is equal to the height of this rectangle, and then the width is the width of the rectangle. We multiply those together to get the area of each of those thin rectangles, and then we add them together. That is the fundamental defining property for the integral, and when we think of an integral, we should think of it as a process of summing products and sums.

So, here we go. I wanted to make a small comment about this notation. The notation has an elongated letter s [$\int$]; it's a long—it's really quite attractive, actually, this long letter s—and this was one of the innovations of Leibniz, again. Leibniz had a flair for good notation, which makes a big difference in our ability to understand it. In particular, in the case of the integral, the big s reminds us of several things. First, it's the first letter of the word sum, and that's not an accident, it was the first letter of the German word *summe*, which meant sum. Therefore, when we see this long letter s, we should think we're going to add up things. Then the question is what are we going to add up? Well, we're going to add up products, and you see that inside the integral side we have velocity, $v(t)$ and then dt. Well, that reminds us that what we do is we look at the velocity at each increment of time, that's the height of one of these rectangles, and then we multiply. This dt was the Δt represents the width of each of those rectangles. So, the product of the width of the rectangle times the velocity—that is the height of the rectangle—gives us the area of that rectangle. And then we're adding them up—that's the s—as the t varies between the lower limit of integration, the 0, and the upper limit of integration, the 3. So, this notation really does capture and remind us of the defining features of the integral.

Now, we're trying to make a graphical correspondence to this method for figuring out the distance traveled, and the graphical correspondence is that this same procedure is telling us the area under the velocity curve. The area under the curve is going to be exactly the same because as we got finer and finer intervals, we see that, in fact, that long sum of products is just approximating the area under the curve. Well, knowing that, we can see that there are alternative ways for figuring out the distance traveled for our car that's moving at $2t$ miles per minute, because we know that the distance traveled is going to be equal to the area under the curve. Well if a curve is really just a straight line with slope 2; it actually just represents a triangle. It's a triangle whose base is 3, whose height is 6, and we know how to compute the area of a triangle. The area of a triangle is just the base times the height divided by 2. Consequently, we know that since the area of this triangle is 9 units, we also know that the distance that the car traveled, if it traveled always at $2t$ miles per minute, would be exactly 9 miles during that 3-minute time interval. And, in fact, instead of thinking of the distance traveled between time 0 and time 3 if we just thought of a variable time, we just chose any time at all, we would be able to compute the area of the triangle very simply because we would know what the height of the triangle was. If, for example, if we had some time t, its height would be $2t$; and, consequently, the width would be

t, the height would be $2t$, and so we would see that the area of the triangle would just be t^2.

In general, then, because the characteristic picture of the integral is obtained by just drawing these approximating rectangles, and as the rectangles get smaller and smaller, the product of their height times their width plus the height times width plus height times width plus height times width is just giving us the sum of the areas of the rectangles. We can see that the integral from the point A to the point B of some function $v(t)dt$ will just give us the area under the curve between A and B. So, that is the graphical representation manifestation of the integral; the interpretation of the integral.

Now, this interpretation leads to many simple observations about the integral. One thing is if we have any function at all, and we take the integral between one point, A, and another point, B, and we add to that the integral from the point B to some further point C, we will obtain the integral from the point A to the point C. Why? Because the integral from A to B of our function is just telling us the area under the curve and above the horizontal line axis of that function; and, then, the integral from B to C is telling us the corresponding area in the next part; and, consequently, adding them together will give us the area under the curve between A and C.

We need to make some comment about times when the function goes below the axis; that is, when the value of the function which we're integrating is negative. In the model of the car moving on the straight road, that means when the car is going backward down the road, it has a velocity in the reverse direction. Then we call that the negative velocity, the velocity would be a negative value, but the method of computing the integral is exactly the same as before; namely, we take the velocity times the small intervals Δt and add them to get the distance—it's the signed distance traveled, sign meaning plus or minus distance traveled. Since you're moving in the negative direction—if the velocity is negative, you're moving backwards on the road—then, you're computing for values where the velocity is negative, the value of the integral will be the negative of the area; it will be the negative of the area between the graph and the horizontal line, if the graph goes below the axis. But, the point is, that if you take a graph where some of the time you're traveling in the positive direction, if this is a velocity curve, every point where the velocity is positive means that you're moving forward along the road. When the velocity is negative, you're moving in the reverse direction. If we compute the net distance traveled from some time A to some time C, we can compute this in the following way: During the time when you're velocity is positive, you'll be moving forward along the road; and, during that time, whatever the area is under that part of the curve will be the distance between where you started and the forward distance that you end up. So, in this case, it's 5 miles if we assume the area under these curve is 5.

Now, we're returning backwards. At this point, the velocity is 0; and when it descends below the axis, that means we are now driving in the reverse direction. And the area under the curve between B and C will accumulate the value of the distance traveled in the reverse direction. So, the total net effect is that if we travel from time A to time C, we will have traveled a total that is a net distance of 3 miles. We actually traveled a total distance of 7 miles, 5 forward and 2 back. So, an integral is not a good way to compute your gas usage if you're going sometimes forward and sometimes back. But, it is a good way to tell the difference between where you started and where you end up. Notice that it maintains the relationship that the integral from A to C is equal to the integral from A to B plus the integral from B to C; this relationship is true no matter where the value or the function is, positive or negative.

So, at this point, we have seen the graphical interpretation of the integral as the area under the curve. And, regardless whether it's a velocity curve or if it's just a general curve, $f(x)$, the integral is defined the same because of the idea of mathematical abstraction. There's no reason that we can't substitute the variables x for t or f for p; the integral is defined in the same way.

Now, I'd like to turn our attention to trying to understand the integral from a different point of view; and, to accomplish that, let's first of all think about the idea of the integral as an accumulation of distance traveled over varying distances, varying times, which we will call x. So, suppose that this is our graph of some velocity function that we are considering. So, this is a velocity function; here we're going forward; here we're standing still for a while, the velocity is 0; down here we're turning in the backwards direction, and so on. The integral from A, this fixed number A, this site, this starting time A. If we measure how far the accumulated, the net distance traveled is for any distance in time, for any time in the future; we're really computing the integral of this velocity function, starting at A and moving forward. So, we have the area of this or the area of this. As our x moves along this axis, the integral of that is going to tell us the net distance traveled. And, if we graph that, we will actually be graphing our position function on the road. In other words, we start here at the beginning. Since this is telling us that the velocity is positive and forward, the integral up to a given point will be some positive value; and, then, here when the velocity is 0, we don't change our position, we're not moving forward or backwards. Here the velocity is negative; consequently, we're moving backward along our road. Then, after we get to this point where the velocity becomes positive, then we begin to move forward on the road again, and that gives us a function. Now, notice that this function is related to the given function that we—the velocity function that we started with; and, this function will actually have its derivative equal to the velocity function. Because, you see, this is a position function determined by the velocity function. So, the velocity function is the derivative of this position function.

Well, we can interpret the graphical meaning of this integral in the following way: If we think of this as a velocity function, then the integral from this fixed point A up to x will have various properties as we think about how the car is moving. So, this is basically summarizing what I just said. That if our velocity function is positive, when it's positive, the accumulated distance will be increasing; the integral is an increasing value. During the times when the function is decreasing, that is, the velocity is going in the backwards direction, then we would expect the accumulated distance, that is to say, the integral, will be decreasing; the integral from a fixed point up to the point x of what you think of this is the velocity function, that it will be decreasing. Likewise, when the velocity function is 0, then the integral stays constant during that time. So, this is showing us that at different points x, whether we would expect, given these velocity properties, what we expect from our accumulated distance, which is our integral.

Now, let's turn to an algebraic representation of this same idea. Suppose that we think about having a velocity function $2t$, and then we say, well, at each point x, what is the distance traveled? Well, the distance traveled is the area under this curve, which is the height, which is $2x$; times the width, which is x; and the area of a triangle is the height times the width divided by 2, which is, if the speed is $2t$, so this is a curve of slope 2, then we find that the integral from 0 to x is the area of that triangle, which is simply x^2.

Now, let me remind you of the Fundamental Theorem of Calculus and how it relates to this issue of the velocity function. Recall that we showed that by taking this sum process associated with the velocity, now we know we get two things when we do that. We get the area under the curve, with the plusses and minuses, depending on whether the curve is above or below the axis, of course; but, it's the area under the curve. But, in addition to it, it is, if this is the velocity function, if we consider this to be the velocity of the car at each moment of time, that integral is going to tell us the net distance traveled between the time A and the time B. Well, we saw the insight of the Fundamental Theorem of Calculus was that if we can find a position function, that is a function that tells us where we are at each moment, with the property that its derivative is equal to the velocity function we're given, then there's this other, simpler way of finding the net distance traveled; namely, just saying well, where were we at time B, and where were we at time A; let's just subtract the two and that will give us the net distance traveled. So, in other words, the area under the curve of a certain function is equal to taking another function whose derivative is what's inside, what you're trying to find the integral of, and subtracting those two values, the value at the top limit of integration minus its value at the bottom limit of integration. That was the Fundamental Theorem of Calculus. It associated the derivative with the integral.

So, the point of the Fundamental Theorem of Calculus is that if our challenge is to find an integral of a function, it suffices to find some other function, in this case capital F, such that the derivative of capital F is the function whose integral we're trying to take. So, our challenge in trying to take an integral, we could say well, what we need to do when we're trying to take an integral is divide up the area between A and B into tiny, tiny parts, and then add up all those little areas, and then take a limit as we take finer and finer divisions of the interval. That is what the integral is telling us. But, that value is equal to the value we get if we can find a function whose derivative is what it is we're taking the integral of.

Now, the point is that this tells us that the burden of taking an integral is solved if we can find what's called an "antiderivative." An antiderivative means a function whose derivative is the function whose integral we're trying to take. We're trying to do the opposite of the derivative process. This, then, is where the algebraic manifestation of integrals is going to appear. If we're trying to take the integral of $2x$ from one place to another, that's what we talked about before, we recognize that x^2 is a function whose derivative is $2x$. And, therefore, if we're trying to take the integral of $2x$ between two points, let's say 1 and 3, knowing that an antiderivative is x^2, we can plug in the upper limit of integration, 3, to get 9; plug in the lower integration, 1, to get 1; and we see that $9 - 1 = 8$ is the integral of $2x$ between 1 and 3.

Let's consider another one. Suppose that we want to take the integral of just x. Well, we find a function whose derivative is x. Here's an example, $\dfrac{x^2}{2}$ is a function whose derivative is x, and we saw that from the algebraic methods of finding derivatives. And, consequently, we are in a position to quickly and easily take the integral of x between any two values.

What's the antiderivative of a constant function? That is, a function whose graph is simply horizontal. Well, an antiderivative is that constant times x, because the derivative of a constant times x is just the constant.

In general, if we are given a function x^n, and our goal is to try to find a function whose derivative is x^n, we have recourse to looking back at our derivative formulas and realizing that if we take the function $\left(\dfrac{x^{n+1}}{n+1} \right)$, and we take it's derivative—remember how we took the derivative; we took the exponent and we bring it down, and then we take x to one lower power, and that $\dfrac{1}{n+1}$ is just a constant, so then the $n+1$'s cancel and we're left with x^n. So, here we are. We see that the derivative of this expression $\dfrac{x^{n+1}}{n+1}$ is exactly x^n; so we found an antiderivative of x^n.

Now, you may note that in each of these cases, I have talked about an antiderivative. Why don't I say the antiderivative? Why don't I say the antiderivative? When I say, "What is an antiderivative of $2x$?" I said, "x^2 is an antiderivative of $2x$, x^2 is a function whose derivative is $2x$." Well, why didn't I say, "The antiderivative of $2x$ is x^2?" Well, the answer is: that if you have any function, if you add a constant to that function, you'll get another function whose derivative is exactly the same at every single value. So, in the case of x^2, for example, if I take $x^2 + 5$, $x^2 + 5$ has exactly the same derivative as x^2 does. It also has derivative $2x$. So, any constant that I add to a function gives me exactly the same derivative because it's just shifted up the slope of the tangent line at every point is exactly correspondent and is exactly the same slope. Therefore, it has exactly the same derivative. That's the geometric way of seeing that it has the same derivative. An algebraic way of seeing that if you have a function and you add a constant, you get the same derivative. It's just realizing that the derivative of the sum of two functions is the sum of the derivatives, and that the derivative of a constant is 0, because a constant is just a horizontal line.

We're now in a position to take our table of derivatives and associate it with every derivative and antiderivative formula. You see, you may recall that our table of derivatives, we saw that when we had the function x had derivative 1; the function x^2 had derivative $2x$; the function x^3 had the derivative $3x^2$; and, in general, x^n had derivative nx^{n-1} power; and we can then make a corresponding antiderivative chart by just looking at those things backwards. How can we manipulate them to say, "If I'm given a function, what is another function whose derivative is what I started with?" And the idea is just to look at this column and take the derivatives and see that what we get is in this column.

Likewise, notice that the derivative of the sine is the cosine; we mentioned that. Therefore, the antiderivative of the cosine is the sine. Since the derivative of the cosine is −sine, the antiderivative of the sine is −cosine; because the derivative of −cosine is the sine. And, in all cases, we can add a constant and continue to get a function whose derivative is as we want it.

The whole purpose of what we've said here is that we are now in a position to do things, such as take an integral of a value, such as what is the area under the graph of the curve x^4 between 1 and 4? That's a curved line. It would be very difficult to get the exact area in any mechanical kind of a way. But, we can take the antiderivative of

x^4, which is $\dfrac{x^5}{5}$; plug in the upper limit of integration, 4; plug in the lower limit of integration, 1; compute this antiderivative at the value 4; subtract the antiderivative at the value 1; and, get the answer. So, this strategy of taking an antiderivative and plugging in the upper limit of integration and the lower limit of integration is what students come to learn as what it means to do integrals.

So, we've seen now a graphical manifestation of the integral and the algebraic manifestation of the integral; namely, antiderivatives. See you next time.

Lecture Eleven
Abstracting the Integral—Pyramids and Dams

Scope:

We saw the power of the derivative in its applications beyond motion: the dynamic view of areas and volumes, the growth of stalactites in caves, and supply and demand curves in economics. Similarly, the integral, when viewed abstractly, is an important tool for understanding diverse dynamical situations such as (again) areas and volumes, as well as engineering. In this lecture, we work out the volume of a pyramid, the volume of a cone, and a solution to an engineering problem: the hydrostatic pressure on a dam.

Outline

I. In this lecture, we will see how the integral can be applied to questions that go beyond a car going on a straight road.

II. Areas and volumes are natural applications of the integral.

A. Consider a square with side length x and area $A(x) = x^2$.

    1. When we studied the derivative, we saw that $A'(x) = 2x$ essentially represents the change in the area when we increase the side length by 1.

    2. This means that for each Δx increase in the side length, the area increases by approximately $2x\Delta x$.

    3. Thus, the total area of a square of side length 5 is the sum of all pieces for each Δx between 0 and 5, and it can be realized as an integral and evaluated by the Fundamental Theorem of Calculus:

$$A(5) = \int_0^5 2x\,dx = 5^2 - 0^2 = 25$$

B. Now consider a cube with side length x and volume $V(x) = x^3$.

    1. When we studied the derivative, we saw that $V'(x) = 3x^2$ essentially represents the change in the volume when we increase the side length by 1, remembering that adding to the length of the side adds a layer of extra volume on three faces of the cube.

2. Thus, again, for each Δx increase in the side length, the volume increases by approximately $3x^2\Delta x$.

3. The total volume of a cube of side length 5 is simply the integral of this function: $V(5) = \int_0^5 3x^2\,dx$

4. We can compute the integral using the Fundamental Theorem of Calculus. Because an antiderivative of $3x^2$ is x^3, we simply evaluate x^3 at the upper limit of integration and subtract x^3 evaluated at the lower limit of integration:

$$V(5) = \int_0^5 3x^2\,dx = 5^3 - 0^3 = 125$$

III. Now, let us look at more complicated examples, such as computing the volume of a pyramid with a square base of side length 200 ft and a height of 200 ft.

A. We can approximate the pyramid by a stack of slightly thickened squares placed on top of one another, in which the squares get smaller as we get near the top of the pyramid.

B. The volume of each thickened square is easy to write down, namely, it is the area of the square times the thickness.

C. The area of a square h units from the top is h^2, so the volume of each slice is approximately $h^2\Delta h$.

D. Adding up the volumes of those thickened squares gives an approximation to the volume of the pyramid. Thus, the total volume of the pyramid is $\int_0^{200} h^2\,dh$.

E. To evaluate the integral, we use the Fundamental Theorem of Calculus. An antiderivative of h^2 is $\dfrac{h^3}{3}$, so

$$\int_0^{200} h^2\,dh = \frac{200^3}{3} - \frac{0^3}{3} = \frac{200^3}{3} = 2{,}666{,}666\text{ft}^3$$

IV. Let's compute the volume of a cone of base radius 3 and height 4.

A. We can view the sideways cone as skewered on the x-axis, and we can think of slicing it up into thin slices as we would do to a loaf of bread.

B. Each slice is approximately the same volume as a slightly thickened disk, and the total volume of the cone is approximately equal to the sum of the volumes of those small slices.

C. Using similar triangles, we can see that at point x, the radius of the disk is $\dfrac{3x^3}{4}$, so the area of the disk is $\pi\left(\dfrac{3x}{4}\right)^2$, or $\dfrac{9\pi}{16}x^2$: thus, the volume of a slice is $\pi\left(\dfrac{3x}{4}\right)^2\Delta x = \dfrac{9\pi}{16}x^2\Delta x$. The total volume of the cone is

$$\int_0^4 \frac{9\pi}{16}x^2\,dx = \frac{9\pi}{16}\left(\frac{4^3}{3} - \frac{0^3}{3}\right) = \frac{36\pi}{3} = 12\pi,$$

or $\dfrac{1}{3}$ times the area of the base $(\pi 3^2)$ times height (4). Again, note the use of the antiderivative of x^2: $\dfrac{x^3}{3}$.

V. The integral is important because the process of summing that the integral is performing is precisely what we need to do to solve various problems in various settings.

A. Suppose we are building a dam and want to know the total hydrostatic force on the face of the dam.

B. The pressure (that is, force per square foot) varies with depth. The pressure is greater near the bottom of the dam than at the top.

C. The total force is obtained by adding up how much force there is at each depth.

D. Suppose the dam is 100 feet wide and 40 feet deep. Then, we could think about dividing the face of the dam into narrow strips of Δh height for the width of the dam.

E. The amount of force on a strip at depth h is the product of the area of the strip, $100\Delta h$, and the pounds per square inch of water pressure at that depth.

F. The water pressure at a depth h is about $62.5h$ pounds per square foot.

G. Thus, the force on a strip of height Δh that lies at depth h is about $(62.5h)(100\Delta h)$ pounds, so that the total force is $\int_0^{40} (62.5h)100\,dh$.

H. We can evaluate this integral via the Fundamental Theorem of Calculus. Given that an antiderivative of h is $\dfrac{h^2}{2}$, we can calculate that the total hydrostatic force on the face of the dam is 5,000,000 pounds.

Readings:

Any standard calculus textbook, section on applications of the integral.

Questions to Consider:

1. How does the integral help us view the area of a circle in a dynamic way?

2. Hooke's Law states that, within certain limits, the force exerted by a spring that is stretched to x units beyond its resting length is a constant times x. Suppose that for a given spring, the spring constant is 5. *Work* in physics is just force times distance. Write an integral whose value equals the work done to stretch this spring from its resting position to one that is 3 units longer. (Note: If a constant force F were applied over 3 units, the work done would simply be $3F$. With the spring, the force varies; thus, $F(x)$ is the variable force where $F(x) = 5x$, and where x is the amount the spring is stretched. The integral is perfectly designed to add up force times distance products as the distance varies from 0 to 3.)

Lecture Eleven—Transcript
Abstracting the Integral—Pyramids and Dams

Welcome back. During the last lecture, we saw the integral interpreted in terms of its graphical interpretation as the area under a curve, and then we also saw how taking antiderivatives gave an algebraic significance to the integral. In this lecture, we're going to be talking about how the integral can be used to apply to questions that are beyond just the idea of a car moving on a straight road. As you may recall, when we were talking about derivatives, we talked about how you could interpret the meaning of the derivative in terms of things like the change in the area of a square or the volume of a cube, and we're going to do the similar thing here in having us look at the interpretation of the integral as a process of summing. And, I'm very anxious that when you think of the integral, you think about its applicability to any kind of a situation that involves dividing things up into little tiny pieces and adding them up. That's really the power of the integral, and it particularly has made manifest in issues about areas and volumes. So, that's where we'll start today.

Let's begin by looking at a square, which we've done before. But, we'll just take this square right here and ask ourselves the question, suppose we were faced with the question of finding the area of the square. Now, I understand that we already know what the area of the square is, so this is a—I'm doing this for the purpose of illustrating the strategy of using the integral rather than getting an answer we don't previously know. So, we know that the area of the square is x^2, but let's see if we can think of it in a dynamic sort of a way.

Suppose we think of this square as evolving, as growing from increasingly large squares that evolve up and grow into bigger and bigger squares. What are we doing in the sense of changing the area of the square? In other words, we're accumulating material to the square; and what amount of material are we accumulating? Well, we're accumulating little pieces like this and like this. We're adding on, so to speak, strips along these two sides, and then other strips along these two sides, and if we thought about adding sort of infinitely thin strips; and, of course, we don't really want to think about infinitely thin, we think about very thin ones, adding up the strip, strip, strip—adding them up as it starts from 0 and it goes to any size that we want. We would accumulate the area of the entire square. So, the idea in our minds is to say, "Ah, the way we get the total value of a square would be to add up this strip plus this strip, and then this strip plus this strip, this strip plus this strip." We can see that adding up that number of strips gives us the square. That gives us a hint that we should be talking about integrals, because integrals are a process of summing, and when we are trying to sum up things in order to get the total value, that's a place where we say, "Ah, use an integral."

So, here is the integral associated with the area of a square of size x. In fact, let's be specific. This is an integral associated with a square of size 5. What is it? Well, all we do is we take the integral from 0 to 5 of $2(x)dx$; $2(x)dx$. Now, why is that? Because we can interpret it visually. We're saying that if we think about this whole square, we can think about it as a strip for every value x between 0 and 5. What are we doing? We're saying take the value $2x$ and thicken it up by a little, tiny width thing, Δx, which turns into dx in the limit. Imagine just a little thin strip of two strips like this, and we're thinking of adding them up to accumulate the area of the square, so that visual meaning of adding up what we see inside the integral sine and thinking of the integral as a process of summing tells us that this integral will, in fact, equal the value of the area of the square. Now, notice that we can actually take this integral. By taking the integral I mean doing what we did in the last lecture; finding an antiderivative—an antiderivative of $2x$ is x^2; plugging in the top value, that's 5; plugging it into x^2 gives 25; subtracting the bottom value, that's 0; giving a total of 25. It tells us an answer that was obvious from the beginning and we knew it already, but we have now an idea of how we can interpret a problem in terms of integrals by dividing up whatever we're after into little bits and adding them up. And that tells us what integral we should take to get the answer we seek.

Okay, let's do another example with a cube. Here's a cube, and once again we're going to do exactly the same thing. We say to ourselves suppose we want to find the volume of a cube of a certain size, like 5; $5 \times 5 \times 5$. Now, of course, we know how to do this; we would just multiply $5 \times 5 \times 5$, but let's think of it instead as an accumulation of layers that make us this cube. In other words, we think of starting at a point, and then adding up a little collection of three boundaries of a certain size—this area plus this area plus this area, and then another layer on top of that—always 3, always 3, always 3; coming out to create a bigger and bigger cube. We can see this in our picture; we can take our cube and divide it up for every value x between 0 and 5, what we're going to do is look at the layer which consists of three slabs that are associated with that number x. The three slabs are this thin layer here on this face, the thin layer on this face, and the thin layer in the back. Each of these has a certain amount of incremental volume.

What is the incremental volume associated with that one particular collection of three slabs, associated with the distance x away from the base point? Well, what's its area? It's $x \times x$, so it's x^2, and then thickened up by Δx. So, we basically have $x \times x$, $\times \Delta x$, $\times 1, 2$, and, in the back, 3. So, the incremental volume is $3x^2\Delta x$, and we're adding those up for as x varies between 0 and 5. So, the integral from 0 to 5 of $3x^2dx$ is going to give us the volume of a cube because we know what the integral means; it's a process of summing. It's summing the $3x^2$'s thickened up by Δx as the x varies between 0 and 5. And, we can see this laminated picture of a cube that fills up this cube of size 5 by increasingly bigger layers of cubes, each one with three sides growing. Then, of course, once again, this is an example of an integral we can actually take. We realize an antiderivative for $3x^2$ is x^3; and, consequently, we know that this integral, which is really obtained by an infinite summing process, that is, you need to do infinitely many because you do a finite number and then you do finer, you do another number, and then you take a limit. But, that laborious process can be computed in a more easy way by finding an antiderivative and then plugging in the upper limit of integration, which is 5; so the antiderivative is x^3, $5 \times 5 \times 5$ is 125; and then we subtract the lower limit of integration, which is just 0, 0^3 is still 0; and we end up with a value of 125 for this integral, which is telling us the volume of the cube.

Okay, so far so good. Let's see if we can do one that may be a little bit—slightly more challenging. Let's consider computing the volume of a pyramid. So, here is a beautiful pyramid, and we can imagine it's like the pyramid at Giza, and we have this beautiful pyramid and our goal is to find what the volume is of this pyramid. Well, the strategy of the integral, the perspective of the integral, is to take what it is you want to know and divide it up into thin laminates, and then, if we can write down a sum that would approximate this volume of the total pyramid by a bunch of layers, then often we can write an integral whose value is exactly equal to the volume of the pyramid that we're trying for. So, let's go ahead and see if we can accomplish this. Let's be specific and have a particular pyramid in mind. Suppose we have a pyramid whose height from the apex down to the base is exactly 200 feet; and, so, the base is 200 feet by 200 feet. So, it's a square base, 200 feet by 200 feet, and the height is exactly 200 feet. What is the volume of that pyramid? How can we conceive of it as an integral problem?

Well, the answer is that we view this pyramid as being a collection of layers of squares thickened up. So, in other words, we go from this smooth-sided pyramid, which is—we, instead, change our view to those step pyramids that come up and straight and then up and straight because those are squares thickened up, square thickened up, square thickened up that become smaller toward the top, bigger toward the bottom, that add up to a volume that approximates the volume of this actual pyramid. Now, just look at it. You see, we can imagine the layers at the top and lower down and lower down and lower down, each having a thickness of Δh. At a given distance down from the top, how big is the square that's contributing to the volume of the whole pyramid but that's at distance h down from the top? Well, let's just think about it. If the total pyramid from the top to the bottom is 200, and the base is a square base of, side length, 200; then, if we go down h feet from the top, we'll be at a level of the pyramid that is $h \times h$. So, that means that this slab, a slab at height h, is going to have an area of $h \times h$—that's how big the pyramid is when we go down h units down from the top and we cut it across. We have a square that is $h \times h$, and it's h high. Well, it's h from the top, but the thickness of this slab is just Δh. We're just thinking of a thin slab. Now, look, we're thinking to ourselves okay, if each slice—as h varies between 0 and the total height of the pyramid, 200, if each slice contributed is $h^2 \times \Delta h$, then we want to add up those incremental pieces of volume so that they accumulate to be the volume of the entire pyramid. So, here we go. That's an integral.

So the total volume is the integral as h varies between 0 and 200 of $h^2 dh$. Whenever you see an integral like this, you must interpret it in your mind as its defining value. That is, that's a sum of products. What products are they that you're adding up? Well, you're saying okay, here's the variable h, dh, and h is going to vary between 0 and

200, the limits of integration, and what am I going to add up? I'm going to add up the value of h, h^2 for every number. I'm going to take the h^2 and then Δh + another $h^2 \Delta h$, as h is varying between 0 all the way up to 200. So, we have this long summation whose value is approximating the integral. And, you can see that that summation corresponds to the volumes of each of these slabs, which are $h^2 \times \Delta h$ in volume.

Okay, now once again, the point is that you can interpret this integral to give you the value you're seeking. It's giving you the value of the volume of this pyramid, but now we can actually compute that integral by using the insights we have from the previous lecture and the Fundamental Theorem of Calculus. Namely, if we want to compute an integral, what we seek is an antiderivative for what's inside. What is an antiderivative of h^2? Well, an antiderivative— remember, we're seeking a function whose derivative is h^2. Well, that function is $\dfrac{h^3}{3}$. And we test ourselves with it. We say, okay, $\dfrac{h^3}{3}$, when we take the derivative of $\dfrac{h^3}{3}$, we bring down the 3; the 3 cancels with the denominator 3; and we're left with h^2.

So, we see that computing the integral between 0 and 200 just amounts to plugging in the 200 into the value of the antiderivative, which is that cubed over 3; and then plugging in the bottom limit of integration, which, in this case, is 0; and we get that this value is the total volume of a pyramid whose height is 200 and whose base is 200 by 200.

So, we're always going to take what it is we're seeking and think about dividing it up into little pieces and adding them together. That is the burden of our song for this lecture. Let's try yet another one.

Here's one, a cone. This is a cone. A cone has a circular base and a point, it comes to a cone; this is a cone; and, we're trying to compute the volume of this cone. Now, I know we've already previously computed the volume of this cone; we know what the answer is. We did that in a previous lecture, but we're going to do it again here to see how it could be done using the concept of the integral. So, the way that we want to think about it is to take this cone and slice it into little thin disks that all are layered on top of each other and accumulate to be the volume of the total cone. Whenever we see something that we're trying to compute the volume of, if we can slice it into little pieces that we add together to accumulate to that volume, we should be thinking to ourselves: integral. That's probably an integral. And, indeed, it is. So, here we go, let's go ahead and think of this.

Here we have a cone, and to be specific, we'll make our cone have specific dimensions so that we can focus our attention on this. We'll imagine our cone to have a height of 4 units—from here to here is 4 units—and then the base of the cone has a radius 3; and then the cone point comes down here and is at the origin of our picture. So, our cone is sitting, from your perspective, sideways here with the x-axis going right through the cone like this. Where this radius is 3, and this height here is 4. Now, how can we imagine this as computing the volume by adding up slices? Well, we say look, for every value of x between 0 and 4, we can imagine cutting the cone at that number x, so here at this point x between 0 and 4 we imagine slicing the cone and approximating the volume of a thin slice by a cylinder, a thin cylinder, whose area is a circle that we get by actually cutting the cone, and then we just thicken it up so that it no longer has a side that's going sideways. Instead, it's once again a step kind of function. It goes straight across for a little bit of distance. So, that little disk, that little disk is a contribution to the volume of the entire cone.

Well, what is that slice volume? Well, at every point x, and we can just draw this point x, so we say okay, suppose I have a value x somewhere between 0 and 4; what is the value of the area of the disk that I obtain by cutting the cylinder exactly at that distance away from the cone point? At exactly x distance? Well, in order to do that, we need to realize that the dimensions of that are—well, the area of it is determined by its radius; it's going to be a circle; and the dimensions of it are determined by its radius; and the radius is going to be told to us because we know the ratio of the entire height of the entire cone to the radius of the base of the cone. Its height is 4 and its radius is 3. So, that means an intermediate place, like height x, what will the radius be at height x? It will be 3/4 of x, just like at distance 4 here, when the height is 4, the radius is 3/4 of 4; 3. So, at an arbitrary point x, the radius is 3/4 of x. That means that the area of the circle that is obtained by cutting the cone right at that point is $\pi \times 3/4\, x^2$. That is the formula for the area of a circle of radius 3/4x.

And, then, what do we do? We take the area of that circle and thicken it up by Δx to get an incremental volume, a contribution to the volume, that is the exact volume of a cylinder whose circular base has radius $3/4x$, and whose height or thickness is Δx. When we have such a concept of the slice volume at each value x, we imagine having such slices for every x value between 0 and 4. That tells us how to make an integral. An integral is, then, the integral as x varies between 0 and 4 of $\frac{9\pi}{16}$ that just means $3x$ of 4^2 times πdx. So, the total volume is the integral as x varies between 0 and 4 of our slice volume added up. So, in other words, inside we have $\pi \times 3/4x^3$; and our Δx turns into a dx, meaning that we take arbitrarily fine divisions, and we take a limit as we take smaller and smaller Δx's. That integral, then, is the total volume of our cylinder. And we can, once again, take an antiderivative by realizing that the antiderivative of x^2 is $\frac{x^3}{3}$, and computing—plugging in the top value 4 and the bottom value 0 to get the total value of the volume of that cylinder—of that cone. So, the point is that if you can take something that you want and divide it up into little pieces that add up to an approximation of it, then taking the integral will often give you the exact answer.

Let's move on now to a different kind of a scenario. This is one where instead of talking about physical objects, areas, or volumes, let's talk about a different kind of a situation that shows the strength of the integral is that it can apply to things even that aren't physical and geometric ideas, like area and volume, or even like speed and distance traveled. Instead, let's think about the hydrostatic force of the water against the side of a dam. That is to say, what is the total force of that water pushing against the dam?

Well, let's think about what that means and how it is that we might go about actually trying to figure out such a thing. The problem with trying to figure out the force of water on a dam is that at the shallow levels, the force of the water against the dam is not as great as it is at the bottom levels. We all know this. When you go diving in a swimming pool, the pressure near the bottom of where you dive is stronger than near the top. You can feel it in your ears and you know that the pressure is stronger. So, the question is, if we're trying to figure out the total force against the wall of the dam, we've got to somehow deal with the idea that the force of the water near the bottom is different from the force of the water near the top. In other words, the pressure is not the same near the top as near the bottom. How are we going to cope with that?

Well, okay, let's just see. What do we think about when we do such a thing? To ground our discussion in a specific case, let's assume that the dam is 40 feet deep and 100 feet wide. So, we have this water over here, and it's pressing against the dam, and our challenge is to find the total hydrostatic force, that's the force by the water, against the side of the dam. Now, one thing we know about water is that at any given depth the force, that is the pressure, the pounds per square inch of that water, is the same at the same depth because it's really just talking about the weight of the water that's above it that's pushing down. That's where the force comes from; it's the weight of all that water that's pushing against it. And, then, something that we need to know from physics is if you are at a certain depth, the water pushes to the side in the same amount that it pushes downward. So, the hydrostatic force against the wall of the dam is just determined by the water above it; the depth of the water.

Well, then, what that means is that if all the water were at the same depth, we would be in a position to figure out what the total force against that part of the dam is, because if it were at the same depth, we would say well, what is the pressure at that depth, and then just multiply by the area. See, the pressure is the pounds per square inch. So, if we had a strip along of the dam at a certain depth, let's say h feet down from the top, we would be able to say well, the force of the water at that depth is so many pounds per square inch, or per square feet, and if we had a certain number of square feet at that depth, then we'd know what the total force is at that depth. So, that's exactly what we do. We say to ourselves, "Ah, wait a minute. Then I know how to do this problem." The way I would do this problem is I would look at a thin strip at a certain depth h. I would imagine myself going down h feet down into the water; I'd take a thin strip of Δh, with the concept in mind that during that thin strip, the difference in the force from the top to the bottom of the strip is not too much because it's more or less at the same depth; not exactly, but more or less.

What would the area of that strip be at depth h? Well, let's see. It would be the area of the strip, which is Δh, times the width, which we assume to be 100 feet, times the pressure, that is, the pounds per square foot of the water at depth h. Well, it turns out water weighs 62.5 pounds approximately per cubic foot, and so at depth h the pressure is $62.5h$ pounds per square foot. So, therefore, that means that $62.5h$ is the pounds per square foot at depth h, and the strip that we're talking about has $100\Delta h$ square feet in it. That's this thin strip here, has $100\Delta h$ square feet in it. So, this product is the force that lies on that thin rectangle. Well, now look. If we add up the force on this rectangle and the force on a rectangle right below it, and right below it, and right below it, all the way down; rectangle, rectangle, rectangle, rectangle, horizontal rectangle all the way to the top, we would get the total force on that dam. Well, we've got to say to ourselves integral; this is an integral. Because, what we're doing is we're saying for every h varying between 0 and 40, what we're doing is taking the width of this strip, which is 100, times Δh, which in the limit becomes dh, because our strips become thinner and thinner, so that variation in pressure, even from the top of the strip to the bottom of the strip, although very small, is not completely negligible, but in the limit we get the exact answer. So, the $100\ \Delta h$ would refer to the area of a strip at depth h. We multiply it by the pounds per square foot at depth h feet, $62.5h$; and that integral from 0 to 40 of $62.5h \times 100dh$ is going to give us the total force on the dam.

Now, we can actually do this integral. We can actually do it because of the fact that we know that 62.5 is a constant, 100 is a constant; we're really just taking the integral of h times a constant. The integral of h is just $\dfrac{h^2}{2}$; we can find an antiderivative. So, we can find that antiderivative. So, we have $62.5 \times 100 \times \dfrac{h^2}{2}$; of course, we can simplify this. And then we plug in 40 to the h; that is, we square it and then subtract what we get when we plug in 0; of course, we get 0. And, then, that computation gives us the total force on the dam.

So, this lecture has been about interpreting the integral as a process of summing. And when we have a problem that can be visualized, and this very often happens, by taking some question and dividing it up into pieces that add up to what it is that we're after, then that's calling out for the integral. In our next lecture, we'll see an application of the integral to probability. I'll see you then.

Lecture Twelve
Buffon's Needle or π from Breadsticks

Scope:

Calculus finds applications in many corners of the world, so it should come as no surprise that calculus is useful in many branches of mathematics, as well. Here, we explore an example in which calculus is used to compute a surprising result in probability. What's especially surprising is that we can compute a definite number, namely, the number π, using a random process. Random processes can lead to unrandom conclusions. In this lecture, we will explore an experiment called *Buffon's Needle*, which involves dropping needles randomly on a sheet of paper. In order to analyze this scenario, we will need to use the sine and the cosine functions, so we begin today's lecture with a review of what the sine and cosine functions are and what their derivatives and integrals are.

Outline

I. In this lecture, we will compute a definite number, namely, the number π, by exploring an experiment called *Buffon's Needle*, which involves dropping needles randomly on a sheet of paper.

II. To solve the Buffon's Needle problem, we will use the Fundamental Theorem of Calculus to compute the integral of the sine function over a certain interval. First, however, we need to get a better understanding of sine.

   A. Recall that sine is a function that is defined geometrically on the circle and that associates a number with every angle. We use radian measurement of the angle to tell us the distance along the unit circle from the point (1,0) counterclockwise up to the point in question.

   B. Sine of the angle θ is the height of a right triangle with angle θ and hypotenuse 1; the cosine is the width of that triangle.

C. As θ changes, so does the sin θ: When θ is 0, sine is 0; when θ is small, so is sin θ; when θ approaches 90 degrees (or $\frac{\pi}{2}$ in radian measurement), sin θ approaches 1.

D. More generally, the (cos θ, sin θ) are the coordinates of the point on the circle of radius 1 corresponding to the angle θ radians.

E. As we rotate our angle around the circle, the sine value varies from 0 up to 1, back to 0, down to −1, back to 0, and so forth. We can graph the sine function and see that it is an oscillating curve.

III. The cosine function is the analogous computation for the horizontal distance to every point on the circle. The cosine measure of angle 0 is 1, and it then oscillates back and forth.

IV. Now let's try to understand the derivative of sine.

    A. Look at the rate at which the line opposite the hypotenuse is changing in relation to a change in the angle.

    B. We ask ourselves how quickly the sine of angle θ will change as we change the angle a small amount.

    C. Notice that the graph of the cosine captures the slopes of the tangent lines on the sine graph. That is a visual indication that the derivative of the sine is the cosine.

    D. We use the fact that the tangent line of a circle is perpendicular to the radius and find similar right triangles in a figure of the unit circle that illustrates the sine function.

    E. We see that the derivative of the sine is the cosine, and the derivative of the cosine is negative the sine.

    F. From the graphs of these functions, we see geometrically why their derivatives are related as they are. Neat.

    G. Now that we have derivatives of sine and cosine, by the Fundamental Theorem of Calculus, we also have their antiderivatives:

        1. An antiderivative of $\cos x$ is $\sin x$.

        2. An antiderivative of $\sin x$ is $-\cos x$.

V. We now look at something entirely different by taking a brief excursion to probability.

A. We can quantitatively describe the chance of an uncertain event.

B. For example, the chance of rolling a 3 when rolling a die is $\frac{1}{6}$.

C. In general, the probability of an event measures what percentage of the time that event will happen.

D. One way to measure probability is to do the experiment many times and just count the fraction.

E. For example, in the die case, we could roll a die many times and see what fraction of the time we get a 3.

F. I did this experiment with the help of my children. They rolled the die 1,000 times and counted 164 3s. The fraction of 3s rolled was actually $\frac{164}{1000}$, which in decimal form is 0.164. This result is close to the probability that we reasoned it must be, $\frac{1}{6}$, which in decimal form is 0.16666…

G. In general, the more times we perform an experiment, the closer the experimental fraction will be to the actual probability. This concept is called the *Law of Large Numbers*.

VI. Now we use probability and calculus to understand Buffon's Needle.

A. The 18th-century French scientist Georges Louis Leclerc, Comte de Buffon, asked a question about a random experiment.

B. The experiment involves dropping needles (or, in our case, breadsticks) on a lined paper.

C. Drop a needle randomly on a lined page where the distance between lines is equal to the needle length. What is the chance that the needle will hit a line?

D. Repeat the process of dropping the needle a number of times and count the times it hits a line. The number of times the needle crosses a line divided by the number of times we dropped the needle is a measure of the frequency with which we hit a line.

E. If we drop the needle more and more, that measure of the frequency should get close to the actual probability.

F. What we will see is that by doing this experiment, we can estimate the value of π.

VII. We can use calculus to deduce what the exact probability should be.

A. Let's describe where the needle could land.

B. There are two parameters we will consider associated with each needle's landing.

    1. One is the angle at which it lands (the angular measurement). If it lands exactly parallel to the parallel lines, its angle is 0. As it rotates, we will measure its position in radial angle from 0 to π.

    2. We will also measure where the center of the needle lands relative to the lines. The center could be on the line or somewhere between the lines.

C. For convenience, we will say that the distance between the lines and the length of the needle are both 2 units.

D. In this way, the position and the angle tell the story.

E. Our challenge is to see how many of those positions hit the line.

F. If the angle is close to 0, then the center must be very close to the line to cause a hit.

G. If the angle is about $\dfrac{\pi}{2}$, then the center can be far away and still cause a hit.

H. Can we make that specific?

I. For any angle θ, if the center's distance is less than $\sin \theta$, the needle will hit the line.

J. Every angle has a particular distance where a needle at that angle first starts to encounter the line.

K. We have, then, a rectangle describing possible positions of the needle.

L. Within that rectangle, those positions under the $\sin\theta$ curve are positions that hit, and those above the curve are positions that don't hit.

M. The total rectangle has area π.

N. The question is: How much area is under the curve?

O. Calculus comes to the rescue.

P. The integral of $\sin\theta$ from 0 to π is 2.

Q. Thus, the probability that the needle will hit a line is $\dfrac{2}{\pi}$.

VIII. This experiment shows a method for estimating the value of π.

    A. We now know that the probability in the abstract of the needle hitting a line is $\dfrac{2}{\pi}$. If after many experiments, we find that a needle hits the line x times in y droppings, then we would expect that $\dfrac{x}{y}$ is about equal to $\dfrac{2}{\pi}$. That is, π is about equal to $\dfrac{2y}{x}$.

    B. Let's see what happens with the data we collected before.
        1. When we dropped the needle 100,000 times, we hit a line 63,639 times.
        2. Using a calculator, we see that $\dfrac{2\times100,000}{63,639} = 3.1427269...$, quite close to π, which is 3.1415926....

    C. Buffon was able to give estimates for π by, we kid you not, throwing breadsticks over his shoulder on a tiled floor and seeing how often they hit the grouting.

    D. Hundreds of years after Buffon tossed his breadsticks, atomic scientists discovered that a similar needle-dropping model seems to accurately predict the chances that a neutron produced by the fission of an atomic nucleus would either be stopped or deflected by another nucleus near it—even nature appears to drop needles.

E. Buffon's Needle used calculus and gives one way of estimating π. Another way to use calculus to estimate π is by doing an infinite addition problem, as we will see later.

F. Perhaps this story could be called: *Randomly dropping a needle from the sky gives us the ability to approximate π.*

Readings:

Any standard calculus textbook, sections on applications of the definite integral.

Burger, Edward B., and Michael Starbird. *The Heart of Mathematics: An invitation to effective thinking.*

Questions to Consider:

1. In what sense do repeated trials of an experiment lead us to conclude the probability of an event happening? Why do more trials result in increasing accuracy?

2. Find a website about Buffon's Needle and try the virtual experiment.

Lecture Twelve—Transcript
Buffon's Needle or π from Breadsticks

Welcome back to *Change in Motion: Calculus Made Clear*. As you know, calculus finds applications in many corners of the world; many applications in science, in economics, and all sorts of things, many of which we'll see in the future lectures and have seen in the past ones. So, it should come as no surprise that calculus also is useful in many branches of mathematics. So in this lecture, we're going to explore an example in which calculus is used to compute a surprising result in probability.

What I find especially surprising is that probability—that is, a random process—can be used to compute a definite number; namely the number π. In other words, a random process can lead to a completely unrandom conclusion. So, in this lecture, we're going to explore an experiment that's called "Buffon's Needle" experiment. This involves dropping needles randomly on a sheet of paper in order to analyze how often those needles are going to hit some lines. But, before we get to that, it turns out that in analyzing that experiment, we need to understand the sine and the cosine functions a little bit. I know we've mentioned them briefly before, but I thought it would be a good idea to begin today's lecture with a review of the sine and the cosine functions, what they are, and what their derivatives and their integrals are. So, we'll begin with the sine and the cosine.

Recall that the sine and the cosine functions are functions that are associated with every angle. The way that the sine function is defined is that we imagine a unit circle. This is a circle centered at the origin of the plane that has radius 1. Every point on that circle, then, is associated with an angle, and that angle is associated with the distance around the circle that we go in order to get to that point. So, we're going to use what's called "radian measurement of angles." So, the radian measurement just tells what the distance is along this unit circle up to the point in question. So, that is to say, an angle of θ means that there is distance θ along this unit circle to get up to this point.

Now, of course, every point in the plane has two coordinates. It has a coordinate on the horizontal direction and a coordinate in the vertical direction. The vertical coordinate is that height of the point at angle θ

is called the sine of θ. Another way to think of the sine of θ is the ratio of the height of a right triangle opposite an angle, the height of the leg opposite the angle divided by the height of the hypotenuse. Since in this case the hypotenuse is length 1, the sine in this case for the unit circle is just the vertical height; that is to say, the *y*-coordinate of the point on that unit circle.

Now, let's just understand how the sine varies as the angle θ varies. First of all, when the sine is 0, when θ is 0—that is to say, when we're talking about the angle that just is 0 length; it just is this point right here on the horizontal axis, well, its *y*-coordinate is 0. So, the sine of 0 is 0. The entire circumference of a circle is—remember, the circumference of a circle is $2 \times \pi \times$ the radius of the circle. Since the radius is 1, the entire circumference is $2 \times \pi$. That means that if we go a quarter of the way around the circle—that is, up to the top—we would have gone $\dfrac{\pi}{2}$ distance along that curved circle. The sine at this point—that is to say, if we have an angle that is of what we think of as 90°, but in radial measurement it's $\dfrac{\pi}{2}$—then it's sine is equal to 1 because the vertical height there is equal to 1. When we continue around the circle, the sine at π is equal to 0, and when we go all the way down to the bottom here, that's the sine of $\dfrac{3\pi}{2}$, then its value is −1 because the vertical distance is in the negative direction—so, it's −1—and when we return to 2π, the sine is once again 0.

Let's look at some other numbers in between. So, for example, an angle whose sine we can actually compute would be accomplished by looking at this figure. Suppose we consider a hexagon inscribed in the circle as you see pictured here. A hexagon has 6 equilateral triangles. They fit neatly around a circle because each angle, remember, of an equilateral triangle is 60 degrees, or in radian measurements, $\dfrac{\pi}{3}$. Well, if we configure the hexagon as you see pictured here, then you can see that half of the length of this—this is an equilateral triangle, so the radius is 1, and this length is also 1; consequently, this distance from the horizontal axis up to this point is equal to 1/2. And the angle is just half of 60°, if we're thinking in degree measurements; or, that is to say, half of that is $\dfrac{\pi}{6}$ radian measurement. So, the sine of $\dfrac{\pi}{6} = \dfrac{1}{2}$. So, that is just an example of a particular angle whose sine we can compute.

The point is, as we rotate our angle around the circle, the value of the sine varies from 0, up to 1, back to 0, down to −1, back to 0, and then continues. If we graph that, we will see that the graph of the sine function is this oscillating curve that is rather attractive, and it oscillates up and down after π radians, and then comes back up after 2π radians, and then just repeats the pattern again.

Likewise, the cosine function is the analogous computation for the horizontal distance to every point on the unit circle. That is to say, it's the first coordinate of every point on the unit circle. So, the cosine measure of angle 0 is 1, and then it proceeds downward and oscillates back and forth, as we see.

Well, what we're interested in doing right now is to compute the derivative of this function. Now, recall, the derivative is measuring the rate of change of the function with the change of the variable. Or, in the graphical terms, it's measuring the slope of the tangent line at each point.

Let's recall the definition of the derivative of any function, the definition of the derivative, recall, was that you took the value of the function—you were trying to compute the derivative at a particular value—in this case, we're trying to compute the derivative of the sine function at a particular angle θ. In other words, we imagine we've gone up some angle θ here, and we're asking ourselves how quickly will the sine of θ change as we change the angle, the radial measurement of the angle, a small amount? What's the rate at which the sine changes given a change in the radial distance? Well, the characteristic difference quotient associated with the definition of the derivative is that we look at the value of the sine of $\theta + \Delta\theta$, thinking of $\Delta\theta$ as a small increment in the variable θ; we subtract sine θ to see the total difference in the sine function for a difference of $\Delta\theta$; and then divide by $\Delta\theta$ to see the rate at which the sine is changing given a change in $\Delta\theta$.

Well, let's look at the picture of what the sine of θ is. It's this vertical distance for an angle θ on a unit circle. Then look at the picture of what the value of the sine of θ + Δθ is; this is Δθ, a tiny additional angle, and the value of sine of θ + Δθ would be the vertical height at this point on the unit circle; that's θ + Δθ around the unit circle. So, the difference between sine of θ + Δθ and sine θ is the vertical height of the change in the vertical height from here to here. In other words, it's the length of this vertical leg of this small triangle.

Now, notice that this small triangle and—recall, when we thought about a circle and looking at a circle up close, we saw that it looked like a straight line. Therefore, we can imagine this hypotenuse to just be the hypotenuse of a straight line of length Δθ. Notice that any curve along a circle is perpendicular to the radius. So that is to say that this angle here is a right angle. Well, there's something interesting about this little, tiny triangle. Namely, its hypotenuse is Δθ; its vertical side is sine θ + Δθ − sine θ. But, let's look at the angle.

This angle is θ. The angle between the tangential line and the radius is a right angle. Consequently, this angle here is also θ; it's the angle between two parallel lines. We have a parallel line here and a horizontal line here; we draw this radius, which is cutting those two lines; so, the opposite interior lines are equal. So, this angle is also θ.

So this little angle inside here is $\frac{\pi}{2} - \theta$; which makes this angle up here θ again. In other words, this little triangle, this little angle way up here, is the same angle as this one, which makes this little right triangle similar to this big right triangle. Well since those angles are similar, we can say that the adjacent leg next to the angle θ divided by the hypotenuse of the small triangle will be the same thing as the adjacent leg divided by the hypotenuse of this big triangle. But, in the small triangle, that ratio is exactly the ratio associated with the derivative. It's the sine of θ + Δθ − sine θ ÷ Δθ. And, yet, it is equal to the similar triangle, the similar ratio, the leg next to the angle θ, that's cosine θ, divided by its radius, which is 1. So, we have this ratio right here which says that this number, which, as Δθ approaches 0 is approaching the derivative of the sine, is simply the cosine. So, what we have proved is that the derivative of the sine is equal to the cosine by this very neat geometric interpretation; and likewise, we can do exactly the same analysis to see that the derivative of the cosine is –sine.

Now, let's look at our graphical representation of the sine and cosine and see that these answers make sense. That is to say, notice that the sine at 0 is going up to the right, and the slope of that line is 1. Well, the value of the cosine at that point is 1. Likewise, as we proceed up on the sine's curve, the slope of the tangent line becomes less until at the very top here at $\frac{\pi}{2}$, the value of the slope of the tangent line is 0.

So, the derivative of the sine at $\frac{\pi}{2}$ = 0; and, lo and behold, the value of the cosine at $\frac{\pi}{2}$ is 0, as predicted. Consequently, we have established that the derivative of the sine is the cosine; similarly, the derivative of the cosine is –sine; and, if we look at it backwards— that is to say, finding antiderivatives—we can conclude that the antiderivative of the cosine is the sine. In other words, the derivative of the sine is the cosine, and an antiderivative of the sine is –cosine because the derivative of the cosine is negative to the sine, so putting a –sine there, the negatives cancel out, and the derivative of –cosine x = sine x.

Okay, this is exactly what we need to analyze our experiment that we're going to now undertake associated with Buffon's Needle experiment. So let me just take one short interlude, now, and tell you a bit about probability and what probability is. Probability is making a measurement of the likelihood of some event occurring, when that event is a random event. The most basic example of this occurs in games of chance. For example, when we take a die, such as this die, it has six sides to it; if we roll the die, then whatever side comes up is as likely as any of the other five sides coming up—if it's a fair die, of course. And, so, we say the probability is 1 out of 6. That's the probability that any particular side will arise.

One aspect of the measurement of probability is that if we repeat an experiment many, many times—for example, we roll a die many, many times—and we see how many times—we record how many times it comes up a 1, a 2, a 3, or a 4. If we roll it many times, we expect that any particular face of that die will arise approximately

1/6 of the time. And, indeed, we expect that if we roll the die many, many times, we would expect the ratio of times it comes up, say, a 3 to approach closer and closer to the probability of its coming up a 3; namely, 1 out of 6 times. Now, I actually did this experiment and—well, it's not quite true that I did this experiment, I had my children do this experiment—of taking a whole bunch of dice and rolling them a lot of times. We rolled them 1000 times. And I had them keep careful track of how many times they got a 1, a 2, and a 3; in particular, I was asking about a 3. I said, "How many times will you get a 3 in rolling these die over and over again?" Actually, we used lots of dice and we'd roll them over and over again. And the answer was, after rolling the dice 1000 times, that 164 times a 3 came up on the face of the die. Well, let's see if that seems about right. You see, our expectation is that roughly 1/6 of the time we should see a 3. Well, 1/6 of 1000 =166 2/3. So, the fact that it came up 164 times was pretty close to the actual expected number of times that it should come up a 3. Now, I do have to confess that if I had done this experiment and they had found that only 50 times it came up a 3, I would not have reported that experiment. So, you'll just have to take my word for it that this is a legitimate experiment.

But, the basic feature of probability I want to emphasize right now is that there's a concept called the Law of Large Numbers, which asserts that if one performs an experiment many, many times, then the successes—in this case, a success is defined as getting a 3—divided by the number of trials, the number of successes divided by the number of trials, as we do the experiment more and more times, that ratio should get closer and closer to the actual probability. So, that's the Law of Large Numbers, and that will come up when we talk about the Buffon Needle experiment.

So, let's turn now to an 18th-century French scientist by the name of Georges Louis Leclerc Comte de Buffon. He asked a question about a random experiment, and here was the experiment he considered:

Suppose that we take a striped—in this case we have a table, and we have two lines on the table, and we take what has come to be known in the literature as a needle, but in this case we have a French breadstick; we have a needle whose length is exactly equal to the distance between these parallel lines.; and you can imagine the parallel lines going on forever on all sides of forever, but the experiment consists of the following thing: We take this needle and randomly throw it on the table, and ask ourselves the following question: Does the needle hit the line? Sometimes it does, one of the lines. Or, sometimes we throw it down and it lands so it doesn't hit any of the lines. He wanted to investigate the question of what is the probability that the breadstick will hit a line. Okay?

Now, it sounds like it's—I don't know how interesting that sounds in itself, but it actually turns out to be quite interesting because what we'll see is that by doing this experiment, we can actually estimate the value of the number π, the ratio of the diameter of a circle to its circumference. We can actually get an approximation of that number π by doing this random experiment. Let's see why.

What we're going to do is to analyze the probability of this randomly thrown needle hitting a line. And, the way that we're going to analyze it is to record all possible ways that this needle could land. Well, first of all, it could land anywhere. So, there are two parameters that we're going to identify associated with the possible positions of its landing. One parameter is the angle at which it lands, and we'll measure the angle in the following way: We'll say that if it lands exactly parallel to our parallel lines, we'll say its angle is 0. Then, as it rotates this way, we'll measure our angle, in radians of course, as we rotate round. So, if it's this angle, we'll say that the angle is $\frac{\pi}{2}$, because the radial measurement from this direction on the horizontal line up to this point is $\frac{\pi}{2}$. And, likewise, we'll continue around—we're rotating from the center of the needle—we rotate around, so that it can go all the way to radial measurement π. So, its radial angle will always be some number between 0 and π. Because we're going at the center, you see, once we get to π, this end is getting closer to 0 again. It's some number between 0 and π. That's the angular measurement.

Then we're also going to ask the question about where the center can land relative to these lines. Well notice that the center is going to be somewhere—it could be right on the line, or it could be the most distant place, which would be halfway between two of the lines. For convenience, we're going to say that the distance between two consecutive lines is exactly 2 units. In other words, we're thinking of our needle as being exactly 2 units, which is also the distance between these parallel lines. We're picking the number 2—I'll tell you in advance—we're picking the number 2 because 1/2 of 2 is 1, and then we're going to be talking about sines, and so it would be convenient for half of this needle to be 1 unit in length.

So, let's consider our needle and see how we're going to analyze the question of under what circumstances the randomly thrown needle hits the line or does not hit the line. Can we record that information? Well, first of all, notice that if the angle is very shallow—for example, suppose the angle is 0—then the only possibility for its hitting the line is if it actually lands on the line because if it's even a little bit off the line, if the center is off the line and its angle is 0— that is to say, horizontal parallel to the line—it won't hit the line.

Whereas, if we have some intermediate angle, such as this one; this looks like about $\frac{\pi}{4}$. Remember, $\frac{\pi}{2}$ is 90°, so $\frac{\pi}{4}$ is 45°. If we have this angle at 45°, then if the center point is 0 distance from the horizontal line, of course, it hits. Likewise, it continues to hit, as we think about that center point landing at a more distant value from the horizontal line. At some point, it just barely hits; and then, after that, it quits hitting. So, let's see if we can record that exact moment where it hits, obviously, when it's 0 distance away, and it continues to hit; as we slide this down, keeping the angle fixed, it'll continue to hit until it just barely hits. Now, can we analyze what that just barely hitting is? When will it just barely hit? Well, let's think.

The center of this breadstick—remember, the center is 1/2 distance from either side because the breadstick is 2 units long, so this is— unit measure, this is distance 1; and, think about the vertical distance here. Well, if this angle here is our measure θ, then this angle is also θ, so this vertical distance here is the sine of θ. So, the time when you will just hit, just barely hit, the horizontal line is when the distance of the center point of the needle away from the line is exactly sine of the angle θ. So here we have this animation that illustrates this very clearly. It's right here at this point where the center of the needle is sine θ distance away from the horizontal line that the needle will cease to hit if we're thinking of moving away from the line.

Now let's see if we can record all of this information in a graphical form. Here is a rectangle that's recording all of the possible landings of the needle in the following way: The angle can be anything between 0 and π; and the distance away—that is, of the center—away from the horizontal line, the nearest horizontal line, is some distance between 0 and 1 unit. So, every point in this rectangle corresponds to one possible landing position of the needle. Let's think about any particular angle θ that the needle could land in, and ask ourselves the question: For what distances away from the nearest line would that center point hit the horizontal line and for what distances would it miss? And the answer is, as we already discussed, that for the distance up to sine of θ that corresponds to a position of the needle in which the needle will hit; and, then, if the distance is bigger then the sine of θ, then the needle will miss. So, in other words, those points in this rectangle of possible landing positions of the needle, those that correspond to the needle hitting one of the horizontal lines is in the white area here, the light area here. Whereas, the area outside are those places where the needle has failed to hit the line.

Now all we need to do, then, to compute the probability of hitting the line is to say, "Well, what's the ratio between the area under the curve—that is, the hits—compared to the total possible area, which is the possible landing sites of the needle?" The total area of the whole rectangle is π wide, 1 high; so, it's π units in area. What is the value of the area that corresponds to a hit, a needle hit? Well, it's the area underneath the sine curve between 0 and π. Now, we're in the domain of calculus. We want to figure out what is the area under the sine curve.

The area under the sine curve is the integral from 0 to π of sine xdx. We saw that in previous lectures. But, now, we saw earlier, that the derivative of the sine is –cosine; and we saw from the previous lecture about the Fundamental Theorem of Calculus that the strategy for computing an integral is to simply plug in the value of the antiderivative at the upper limit of integration, so it's –cosine of π, and then subtract what we get by plugging in the antiderivative at the lower limit of integration. So, negative –cosine of π minus –cosine of 0—well, the cosine of π is –1; minus that is +1; negative negative –1—two minuses in a row—gives another +1; 1 + 1 = 2. So, the area underneath the sine curve between 0 and π is exactly equal to 2. That's rather remarkable.

So, that means that the probability of our needle hitting a line is $\frac{2}{\pi}$ because the area of the hits is 2 and the total area of all the possible landings is π. So, the answer is the probability of this randomly thrown needle is $\frac{2}{\pi}$.

Now we return to the idea that when you do an experiment many, many times, the actually number of hits divided by the total number of experiments you do is going to approximate the probability. But, in this case, we've actually computed the probability on theoretical grounds. So, we know the probability is $\frac{2}{\pi}$. So if we actually do the experiment many, many times, we can expect that the number of hits divided by the number of throws will approximate to $\frac{2}{\pi}$. Just doing the cross multiplication, that means that π will be approximately equal to 2 times the number of throws divided by the number of hits.

Well one can look on the computer, on the Web, and actually perform this experiment—simulate the performance of this experiment—many, many times. So, I went to one of these websites and performed it 100,000 times. In that 100,000 times, 63,639 times the needle hit the line. We know that that is an approximation to $\frac{2}{\pi}$. Consequently, we can guess that π is approximately equal to 2 × 100,000 ÷ 63,639l, which is 3.1427269…. Well, π is actually 3.1415926…. You can see that we actually got π to two decimal places by simulating this random experiment.

Now I was told that Buffon himself actually performed this experiment by taking breadsticks and throwing the breadsticks over his shoulder on the floor and, counting the number of times, actually estimated the value of π. Now, I have trouble believing this because I think the breadsticks would break and you'd have to do so many experiments to get close that I don't think that he could actually do that. But, in fact, hundreds of years now after Buffon tossed his breadsticks, or didn't, atomic scientists have recently discovered that a similar needle-dropping kind of model seems to accurately predict the chances that a neutron produced by the fission of an atomic nucleus would either be stopped or deflected by another nucleus near it. So, it appears that even nature appears to drop needles.

Anyway, we've seen from this whole discussion today that Buffon's Needle experiment is an interesting way to apply calculus to something that seems to be completely unrelated—in this case, estimating the value of π. I think maybe this whole story could be summarized by saying *randomly dropping a needle from the sky gives us the ability to approximate π.*

In the next lecture, we'll begin to explore some of the foundational concepts that underlie the mathematics of calculus; namely, the limit, continuity, and differentiability. I'll see you then.

Timeline

All dates are approximate in the sense that the mathematical activities mentioned each spanned several or many years.

540 B.C. ....Pythagoras founded his school and proved the Pythagorean Theorem.

450 B.C. ....Zeno posed his paradoxes of motion.

355 B.C. ....Eudoxus was associated with the method of exhaustion that was an integral-like process.

300 B.C. ....Euclid presented the axiomatic method in geometry in his *Elements*.

225 B.C. ....Archimedes used integral-like procedures to find formulas for various areas and volumes of geometrical figures.

225 B.C. ....Apollonius described the geometry of conic sections.

A.D. 1545...................................Tartaglia, Cardano, and Ferrari were involved with the algebraic solution of cubic and quartic equations.

1600 ....Kepler and Galileo did work on motion and planetary motion, describing those mathematically.

1629 ....Fermat developed methods of finding *maxima* and *minima* using infinitesimal methods resembling the derivative.

1635 ....Cavalieri developed a method of indivisibles.

1650 ....Descartes developed connections between geometry and algebra and invented methods for finding tangents to curves.

1665–1666	During these plague years, Newton devised calculus, his laws of motion, the universal law of gravitation, and works on optics.
1669	Barrow formulated ideas leading to the Fundamental Theorem of Calculus and resigned his chair at Cambridge in favor of Newton.
1672	Leibniz independently discovered calculus and devised notation commonly used now.
1700	The Bernoullis were involved with the development and application of calculus on the continent.
1715	Brook Taylor and Colin Maclaurin developed ideas of approximating functions by infinite series.
1750	Euler developed a tremendous amount of mathematics, including applications and extensions of calculus, especially infinite series.
1788	Lagrange developed ideas about infinite series and the calculus of variations.
1805	Laplace worked on partial differential equations and applications of calculus to probability theory.
1822	Fourier invented a method of approximating functions using trigonometric series.
1827	Cauchy developed ideas in infinite series and complex variable theory and refined the definitions of limit and continuity.

1851 .... Riemann developed a modern definition of the integral.

1854 .... Weierstrass formulated the rigorous definition of the limit used today.

1700–present day .... Innumerable mathematicians, scientists, engineers, and others have applied calculus to all areas of mathematics, science, economics, technology, and many other fields. Mathematicians continue to develop new mathematics based on the ideas of calculus.

Glossary

Acceleration: Rate of change of velocity; a measure of how fast the velocity is changing. The second derivative of position. Units are distance/time^2.

Antiderivative (of a function f): A function with a derivative equal to f.

Brachistochrone: A curve traced between two points (not atop one another), along which a freely falling object will reach the bottom point in the least amount of time.

Continuous function: A function that has no breaks or gaps in its graph; the graph of a continuous function can be drawn without lifting the pen.

Cosine: A function of angle θ giving the ratio of the length of the adjacent side to the length of the hypotenuse of a right triangle, as well as the horizontal coordinate of a point on the circle of radius 1 corresponding to the angle θ.

Delta (Δ): The Greek letter capital delta is used in such expressions as Δt or Δx to denote a small change in the varying quantity. We should think of the Δ as shorthand for "difference." In the definition of the integral, Δx or Δt appears in the long sums used to define the integral. The Δ is transformed into a dx or dt in the integral symbol to remind us of the origins of the integral as a sum.

Derivative: Mathematical description of (instantaneous) rate of change of a function. Characterized geometrically as the slope of the tangent line to the graph of the function. The derivative of a function $f(x)$ is written $f'(x)$ or $\dfrac{d}{dx}(f(x))$ and is formally defined as $\lim\limits_{\Delta x \to 0} \dfrac{f(x + \Delta x) - f(x)}{\Delta x}$.

Differentiable function: A function whose derivative exists at every point where the function is defined; a continuous function without kinks or cusps.

Differential equation: An equation involving a function and its derivatives; a solution of a differential equation is a family of functions.

Directional derivative: The rate of change of a function of several variables in the direction of a given vector.

Direction field: A two-dimensional field of arrows indicating the slope of the tangent line at each given point for a curve that is a solution of a differential equation.

Epicycles: A circle on a circle. For centuries before Kepler, people believed that planets' orbits were circular, but because that image did not accord with observation, the planets were viewed as revolving around on little circles whose centers were going in circles. The smaller circles with centers on circles are epicycles.

Exhaustion (the Greek method of exhaustion): A geometric technique by which formulas for areas of different shapes could be verified through finer and finer approximations.

Fourier Series: An infinite series of sines and cosines, typically used to approximate a function.

Function: Mathematical description of dependency. A rule or correspondence that provides exactly one output value for each input value. Often written algebraically as $f(x) =$ formula involving x (for example, $f(x) = x^2$).

Fundamental Theorem of Calculus: The most important theorem in calculus. Demonstrates the reciprocal relationship between the derivative and the integral.

Graph: A geometric representation of a function, showing correspondences via pairs of points (input, output) drawn on a standard Cartesian (x-y) plane.

Heat equation: A partial differential equation governing the heating and cooling of objects; it relates the second derivative of position to the first derivative of time.

Infinite series (also called an infinite sum): The sum of an infinite collection of numbers. Such a series can sum to a finite number or "diverge to infinity"; for example, $\frac{1}{2}+\frac{1}{4}+\frac{1}{8}+\frac{1}{16}+\ldots$ sums to 1, but $1+\frac{1}{2}+\frac{1}{3}+\frac{1}{4}+\ldots$ does not sum to a finite number.

Integer: A whole number (positive, zero, or negative); $\ldots-2, -1, 0, 1, 2,\ldots$

Integral: Denoted $\int_a^b v(t)dt$. If we think of function $v(t)$ as measuring the velocity of a moving car at each time t, then the integral is a number that is equal to the distance traveled, because the integral is obtained by dividing the time from a to b into small increments and approximating the distance traveled by assuming that the car went at a steady speed during each of those small increments of time. By taking increasingly smaller increments of time, approximations converge to a single answer, the integral. This sounds complicated, but the naturality of it is the topic of Lecture Three. The integral is also equal to the area under the graph of $v(t)$ and above the t-axis. The integral is related to the derivative (as an inverse procedure) via the Fundamental Theorem of Calculus. See also **antiderivative**.

Law of Large Numbers: The theorem that the ratio of successes to trials in a random process will converge to the probability of success as increasingly many trials are undertaken.

Limit: The result of an infinite process that converges to a single answer. Example: the sequence of numbers $1,\frac{1}{2},\frac{1}{3},\frac{1}{4},\frac{1}{5},\ldots$ converges to the number 0.

Maximum: The largest value of the outputs of a function. The y-value of the highest point on the graph of a function. It does not always exist.

Minimum: The smallest value of the outputs of a function. The y-value of the lowest point on the graph of a function. It does not always exist.

Newton-Raphson Method: An iterative technique for finding solutions of an equation using graphs and derivatives.

Parabola: A conic section defined as the set of all points equidistant between a point and a line.

Paradox: Two compelling arguments about the same situation that lead to two opposite views. Zeno's paradoxes of motion give logical reasons why motion cannot occur. On the other hand, we experience motion. The opposite conclusions deduced from Zeno's logic versus our experience compose the paradox.

Partial derivative: The rate of change of a quantity relative to the change of one of several quantities that are influencing its value when the other varying quantities remain fixed.

Partial differential equation: An equation involving a function of several variables and its partial derivatives.

π (pi): Greek letter denoting the value 3.1415926... equal to the ratio of the circumference of a circle to its diameter.

Probability: The quantitative study of uncertainty.

Real number: Any decimal number.

Sine: A function of angle θ giving the ratio of the length of the opposite side to the length of the hypotenuse of a right triangle, as well as the vertical coordinate of a point on the circle of radius 1 corresponding to the angle θ.

Slope (of a straight line): The ratio of distance ascended to distance traversed, sometimes known as "rise over run."

Smooth function: A function that is continuous and whose first derivative, second derivative, and so forth are all continuous.

Tangent line: A straight line associated to each point on a curve. Just grazing the curve, the tangent line "parallels" the curve at a point.

Variable: The independent quantity in a functional relationship. For example, if position is a function of time, time is the variable.

Vector: An arrow indicating direction and magnitude (usually of motion in two-dimensional or three-dimensional space).

Vector field: A field of arrows associating a vector to each point (x,y) in the two-dimensional plane; usually represented graphically.

Velocity: Average velocity is total distance divided by the time it took to traverse that distance; units are length/time (for example, miles per hour). Instantaneous velocity is the speed at one moment of time, approximated by average velocity for smaller and smaller time intervals; units are also length/time. The instantaneous velocity is the derivative of the position function for a moving object.

Wave equation: A partial differential equation governing the displacement of a string as well as more complex phenomena, such as electromagnetic waves; it relates the second derivative of position to the second derivative of time.

Biographical Notes

Archimedes (c. 287–212 B.C.). Ancient Greek mathematician, physicist, astronomer, inventor, and prolific author of scientific treatises. He studied hydrostatics and mechanics and discovered the general principle of the lever, how to compute tangents to spirals, the volume and surface area of spheres, the volume of solids of revolution, many applications of the method of exhaustion, and an approximation of the value of π, among other work. Archimedes was killed by a Roman soldier when Syracuse was conquered during the Second Punic War.

Barrow, Isaac (1630–1677). Lucasian professor of mathematics at Cambridge. In 1669, Barrow resigned from his chair to give Newton the professorship. He contributed to the development of integral calculus, particularly through the recognition of its inverse relationship with the tangent. He published works in optics and geometry and edited the works of ancient Greek mathematicians, including Euclid and Archimedes.

Bernoulli, Daniel (1700–1782). Swiss professor of mathematics at St. Petersburg and at Basel. He is best known for work in fluid dynamics (the *Bernoulli principle* is named for him) and is also known for work in probability. He was the son of Jean Bernoulli.

Bernoulli, Jacques (often called Jakob or James) (1654–1705). Professor of mathematics at Basel and a student of Leibniz. He studied infinite series and was the first to publish on the use of polar coordinates (the *lemniscate of Bernoulli* is named for him). He formulated the Law of Large Numbers in probability theory and wrote an influential treatise on the subject. Together, Jacques and brother Jean were primarily responsible for disseminating Leibniz's calculus throughout Europe.

Bernoulli, Jean (often called Johannes or John) (1667–1748). Swiss mathematician. He was professor of mathematics at Groningen (Holland) and Basel (after the death of his brother Jacques). He was a student of Leibniz and applied techniques of calculus to many problems in geometry and mechanics. He proposed the famous Brachistochrone problem as a challenge to other mathematicians. Jean Bernoulli was the teacher of Euler and L'Hôpital (who provided Jean a regular salary in return for mathematical discoveries, including the well-known *L'Hôpital's Rule*).

Buffon, Georges Louis Leclerc, Comte de (1707–1788). French naturalist and author of *Histoire naturelle*. He translated Newton's *Method of Fluxions* into French. He formulated the Buffon's Needle problem, linking the study of probability to geometric techniques.

Cauchy, Augustin Louis (1789–1857). Prolific French mathematician and engineer. He was professor in the Ecole Polytechnique and professor of mathematical physics at Turin. He worked in number theory, algebra, astronomy, mechanics, optics, and elasticity theory and made great contributions to analysis (particularly the study of infinite series and of complex variable theory) and the calculus of variations. He improved the foundations of calculus by refining the definitions of limit and continuity.

Cavalieri, Bonaventura (1598–1647). Italian mathematician; professor at Bologna; student of Galileo. He developed the method of indivisibles that provided a transition between the Greek method of exhaustion and the modern methods of integration of Newton and Leibniz. He applied his method to solve a majority of the problems posed by Kepler.

Descartes, René (1596–1650). French mathematician and philosopher. He served in various military campaigns and tutored Princess Elizabeth (daughter of Frederick V) and Queen Christina of Sweden. Descartes developed crucial theoretical links between algebra and geometry and his own method of constructing tangents to curves. He made substantial contributions to the development of analytic geometry.

Euler, Leonhard (1707–1783). Swiss mathematician and scientist. Euler was the student of Jean Bernoulli. He was professor of medicine and physiology and later became a professor of mathematics at St. Petersburg. Euler is the most prolific mathematical author of all time, writing on mathematics, acoustics, engineering, mechanics, and astronomy. He introduced standardized notations, many now in modern use, and contributed unique ideas to all areas of analysis, especially in the study of infinite series. He lost nearly all his sight by 1771 and was the father of 13 children.

Fermat, Pierre de (1601–1665). French lawyer and judge in Toulouse; enormously talented amateur mathematician. Fermat worked in number theory, geometry, analysis, and algebra and was the first developer of analytic geometry, including the discovery of equations of lines, circles, ellipses, parabolas, and hyperbolas. He wrote *Introduction to Plane and Solid Loci* and formulated the famed *Fermat's Last Theorem* as a note in the margin of his copy of Bachet's *Diophantus*. He developed a procedure for finding maxima and minima of functions through infinitesimal analysis, essentially by the limit definition of derivative, and applied this technique to many problems, including analyzing the refraction of light.

Fourier, Jean Baptiste Joseph (1768–1830). French mathematical physicist and professor in the Ecole Polytechnique. Fourier accompanied Napoleon on his campaign to Egypt, was appointed secretary of Napoleon's Institute of Egypt, and served as prefect of Grenoble. He carried out extensive studies in heat propagation, which form the foundation of modern partial differential equations with boundary conditions. He developed the *Fourier Series*, which represents functions by infinite (trigonometric) series.

Galilei, Galileo (1564–1642). Italian mathematician and philosopher; professor of mathematics at Pisa and at Padua. He invented the telescope (after hearing of such a device) and made many astronomical discoveries, including the existence of the rings of Saturn. He established the first law of motion, laws of falling bodies, and the fact that projectiles move in parabolic curves. Galileo made great contributions to the study of dynamics, leading to consideration of infinitesimals (eventually formalized in the theory of calculus). He advocated the Copernican heliocentric model of the solar system and was subsequently placed under house arrest by the Inquisition.

Gauss, Carl Friedrich (1777–1855). German mathematician; commonly considered the world's greatest mathematician, hence known as the Prince of Mathematicians. He was professor of astronomy and director of the observatory at Göttingen. Gauss provided the first complete proof of the Fundamental Theorem of Algebra and made substantial contributions to geometry, algebra, number theory, and applied mathematics. He established mathematical rigor as the standard of proof. His work on the differential geometry of curved surfaces formed an essential base for Einstein's general theory of relativity.

Green, George (1793–1841). Most famous for his theorem known as *Green's Theorem*. He worked in his father's bakery for most of his life and taught himself mathematics from books. He proved his famous theorem in a privately published book that he wrote to describe electricity and magnetism. Green did not attend college until he was 40. He had seven children (all with the same woman, Jane Smith) but never married. He never knew the importance of his work, but it and its consequences have been described as "…leading to the mathematical theories of electricity underlying 20th-century industry."

Kepler, Johannes (1571–1630). German astronomer and mathematician; mathematician and astrologer to Emperor Rudolph II (in Prague). Kepler assisted Tycho Brahe (the Danish astronomer) in compiling the best collection of astronomical observations in the pre-telescope era. He developed three laws of planetary motion and made the first attempt to justify them mathematically. They were later shown to be a consequence of the universal law of gravitation by Newton, applying the new techniques of calculus.

Lagrange, Joseph Louis (1736–1813). French mathematician; professor at the Royal Artillery School in Turin, at the Ecole Normale and the Ecole Polytechnique in France, and at the Berlin Academy of Sciences. He studied algebra, number theory, and differential equations and unified the theory of general mechanics. He developed classical results in the theory of infinite series and contributed to the analytical foundation of the calculus of variations.

Laplace, Pierre Simon de (1749–1827). French mathematician and astronomer; professor at the Ecole Normale and the Ecole Polytechnique. Laplace was the author of the influential *Mécanique Céleste* that summarized all contributions to the theory of gravitation (without credit to its contributors). He developed potential theory, important in the study of physics, and partial differential equations. He made great contributions to probability theory based on techniques from calculus.

Leibniz, Gottfried Wilhelm von (1646–1716). German diplomat, logician, politician, philosopher, linguist, and mathematician; president of the Berlin Academy. Leibniz is regarded, with Newton, as a co-inventor of calculus. He was the first to publish a theory of calculus. Leibniz's notation is used currently. He made substantial contributions to formal logic, leading to the establishment of symbolic logic as a field of study. He discovered an infinite series formula for $\dfrac{\pi}{4}$. He was accused of plagiarism by British partisans of Newton, and his supporters counterclaimed that Newton was the plagiarist. Now, he is acknowledged to have independently discovered calculus.

L'Hôpital, Guillaume François Antoine (1661–1704). French marquis, amateur mathematician, and student of Jean Bernoulli. L'Hôpital provided one of the five submitted solutions to Bernoulli's Brachistochrone problem. He was the author of the first calculus textbook (1696), written in the vernacular and based primarily on the work of Jean Bernoulli. This text went through several editions and greatly aided the spread of Leibniz's calculus on the Continent.

Lotka, Alfred J. (1880–1949). An American biophysicist and the father of mathematical biology; he published the first book in this field in 1924. With Vito Volterra, he is chiefly known for the formulation of the Lotka-Volterra equations for the study of predator-prey models in population dynamics.

Newton, Sir Isaac (1642–1727). Great English mathematician and scientist; Lucasian professor of mathematics at Cambridge. Newton was the first discoverer of differential and integral calculus. He formulated the law of universal gravitation and his three laws of motion, upon which classical physics is based. In 1687, he published his results in *Philosophiae Naturalis Principia Mathematica*. He formulated the theory of colors (in optics) and proved the binomial theorem. He is possibly the greatest genius of all time. Newton was a Member of Parliament (Cambridge), long-time president of the Royal Society, and Master of the Mint. The controversy with Leibniz over attribution of the discovery of calculus poisoned relations between British and Continental scientists, leading to the isolation of British mathematicians for much of the 18th century.

Riemann, Georg Friedrich Bernhard (1826–1866). German mathematician; professor of mathematics at Göttingen. He made great contributions to analysis, geometry, and number theory and both extended the theory of representing a function by its Fourier series and established the foundations of complex variable theory. Riemann developed the concept and theory of the Riemann integral (as taught in standard college calculus courses) and pioneered the study of the theory of functions of a real variable. He gave the most famous job talk in the history of mathematics, in which he provided a mathematical generalization of all known geometries, a field now called Riemannian geometry.

Weierstrass, Karl (1815–1897). German mathematician. He left the University of Bonn without a degree and taught secondary school for more than 12 years while independently studying analysis. His 1854 paper led to a professorship in Berlin. He provided a rigorous definition of the limit, thus placing calculus at last on solid mathematical ground, and he provided a precise definition of real numbers.

Zeno of Elea (c. 495–430 B.C.). Ancient Greek dialectician and logician. He is noted for his four paradoxes of motion. He was a student of Parmenides, whose school of philosophy rivaled that of the Pythagoreans.

Bibliography

Readings

Bardi, Jason Socrates. *The Calculus Wars: Newton, Leibniz, and the Greatest Mathematical Clash of All Time.* New York: Thunder's Mouth Press, 2006. This book tells the story of the controversy between Newton and Leibniz and their supporters over who should receive credit for the discovery or invention of calculus. The fact that a book on this topic for the general reader is published in 2006 is a testament to both the significance of calculus and the level of rancor of the dispute.

Bell, E. T. *Men of Mathematics.* New York: Simon & Schuster, 1937. A classic of mathematics history, filled with quotes and stories (often apocryphal) of famous mathematicians.

Berlinski, David. *A Tour of the Calculus.* New York: Pantheon Books, 1995. This book is written in a flowery manner and gives the nonmathematician a journey through the ideas of calculus.

Blatner, David. *The Joy of π.* New York: Walker Publishing Company, 1997. This fun little paperback is filled with gems and details about the history of the irrational number π.

Boyer, Carl B. *The History of the Calculus and Its Conceptual Development.* Mineola, NY: Dover Publications, 1959. A scholarly history of calculus from ancient times through the 19th century.

————. *A History of Mathematics.* Princeton: Princeton University Press, 1968. An extensive survey of the history of mathematics from earliest recorded history through the early 20th century. Each chapter includes a good bibliography and nice exercises.

Burger, Edward B., and Michael Starbird. *The Heart of Mathematics: An Invitation to effective thinking.* Emeryville, CA: Key College Publishing, 2000. This award-winning book presents deep and fascinating mathematical ideas in a lively, accessible, and readable way. The review in the June–July 2001 issue of the *American Mathematical Monthly* says, "This is very possibly the best 'mathematics for the non-mathematician' book that I have seen—and that includes popular (non-textbook) books that one would find in a general bookstore."

————. *Coincidences, Chaos, and All That Math Jazz: Making Light of Weighty Ideas.* New York: W.W. Norton & Co., 2005. This book fuses a professor's understanding of rigorous mathematical ideas with the distorted sensibility of a stand-up comedian. It covers many beautiful topics in mathematics that are only touched on in this course, such as probability, chaos, and infinity. "Informative, intelligent, and refreshingly irreverent," in the words of author Ian Stewart.

Cajori, Florian. *A History of Mathematics*, 5th ed. New York: Chelsea Publishing Co., 1991 (1st ed., 1893). A survey of the development of mathematics and the lives of mathematicians from ancient times through the end of World War I.

———. "History of Zeno's Arguments on Motion." *American Mathematical Monthly*. Vol. 22, Nos. 1–9 (1915). Cajori presents a philosophical and mathematical discussion of the meaning of Zeno's four paradoxes of motion.

Calinger, Ronald. *A Contextual History of Mathematics*. Upper Saddle River, NJ: Prentice-Hall, 1999. This modern, readable text offers a survey of mathematics from the origin of numbers through the development of calculus and classical probability. It includes a nice section on the Bernoulli brothers.

Churchill, Winston Spencer. *My Early Life: A Roving Commission* (available through out-of-print bookstores only). Winston Churchill won the Nobel Prize in literature. The writing in this autobiography of his early life is absolutely delightful. We refer to only a few pages about his struggles with mathematics, but the whole book is a joy.

Davis, Donald M. *The Nature and Power of Mathematics*. Princeton: Princeton University Press, 1993. This wide-ranging book does not study the history of calculus in particular; rather, it describes an array of ideas from all areas of mathematics. It includes brief biographies of Gauss and Kepler, among other mathematicians.

Dunham, William. *Journey through Genius: The Great Theorems of Mathematics*. New York: John Wiley & Sons, 1990. Each of this book's 12 chapters covers a great idea or theorem and includes a brief history of the mathematicians who worked on that idea. Mathematicians discussed include Archimedes, Newton, the Bernoullis, and Euler.

Eves, Howard. *Great Moments in Mathematics (after 1650)*. Washington, DC: The Mathematical Association of America, 1981. This collection of lectures includes four entertaining chapters relevant to the development of calculus. These begin with the invention of differential calculus and conclude with a discussion of Fourier series.

Goodstein, David L., Judith R. Goodstein, and R. P. Feynman, *Feynman's Lost Lecture: The Motion of Planets around the Sun*. New York: W.W. Norton & Co., 1996. This book and CD give Feynman's geometric explanation for elliptical orbits of planets, as well as a history of the problem from Copernicus and Kepler to the present day.

Kline, Morris. *Mathematics: A Cultural Approach*. Reading, MA: Addison-Wesley Publishing Co., 1962. This survey of mathematics presents its topics in both historical and cultural settings, relating the ideas to the contexts in which they developed.

Priestley, William M. *Calculus: An Historical Approach*. New York, Heidelberg, and Berlin: Springer-Verlag, 1979. Develops the standard theory of calculus through discussions of its historical growth, emphasizing the history of ideas rather than the history of events.

Schey, H. M. *Div, Grad, Curl, and All That: An Informal Text on Vector Calculus*, 4th ed. New York: W.W. Norton, 2005. This book has been a favorite introduction to ideas of vector calculus for more than 30 years. It is short and clearly written. It presents vector calculus in the context of electrostatics, so it is especially attractive to people who have a feel for the physics of electricity and magnetism.

Simmons, George F. *Calculus with Analytic Geometry*. New York: McGraw-Hill, 1985. This college-level mathematics text provides a standard development of calculus along with appendices that include biographical notes and supplementary topics.

Thompson, Sylvanus P., and Martin Gardner. *Calculus Made Easy*, New York: St. Martin's Press, 1998. This book is a revision, by the great mathematical expositor Martin Gardner, of a classical exposition of calculus for the general public.

Standard Textbooks

There are dozens of standard calculus and differential equations textbooks, usually titled *Calculus* and *Differential Equations*. Two of them are:

Boyce, William E., and Richard C. DiPrima. *Elementary Differential Equations and Boundary Value Problems*, 8th ed. New York: John

Wiley & Sons, 2005. This is the bestselling introductory differential equations textbook. It presents differential equations from the applied mathematicians' point of view and includes explanations of both the theory and applications, as well as many examples, applications, and exercises.

Stewart, James. *Calculus*, 5th ed. Belmont, CA: Brooks/Cole, 2003. The Stewart textbook is the best-selling calculus textbook in the United States. It is well-written and comprehensive and contains many worked examples, applications, and exercises.

Internet Resources

A History of the Calculus. School of Mathematics and Statistics, University of St. Andrews, Scotland. www-history.mcs.st-andrews.ac.uk/ history/HistTopics/ The_rise_of_calculus.html. This site presents a synopsis of the history of calculus from ancient times through the time of Newton and Leibniz and includes links to many more history sites.

Index of Biographies. School of Mathematics and Statistics, University of St. Andrews, Scotland. www-history.mcs.st-andrews.ac.uk/~history/BiogIndex.html. This site contains biographical articles on many of the world's mathematicians from ancient times to the present. Both chronological and alphabetical indexes are presented, as well as such categories as famous curves, history topics, and so forth.

The Math Forum @ Drexel. Drexel School of Education, Drexel University. www.mathforum.org/library/topics/svcalc/. Includes links to many other sites that contain articles and demonstrations of concepts in calculus.

Mnatsakanian, Mamikon. *Visual Calculus by Mamikon*. California Institute of Technology. www.its.caltech.edu/~mamikon/calculus.html. This site offers animations demonstrating some of Mnatsakanian's applications of his clever sweeping tangent method for solving calculus problems.

Predator-Prey Models. Department of Mathematics, Duke University. www.math.duke.edu/education/webfeats/Word2HTML/Predator.htm l. This site provides an interactive location for creating direction fields and solutions to differential equation predator-prey models.

Weisstein, Eric. *Wolfram MathWorld, The Web's Most Extensive Mathematics Resource.* mathworld.wolfram.com. This website is like a mathematical encyclopedia. If you come across a mathematical term or concept, no matter how trivial or how involved, it's likely to be described here.